Table of Contents

ACKNOWLEDGMENTS

Special thanks to S. Joseph Domina, Vice President, Chase Manhattan Bank, the Teleport Communications Group, Inc., and Joan Butterworth.

Your comments and recommendations are welcomed. Please send all correspondence to:

Telstar Resource Group, Inc
15 West 39th Street
16th Floor
New York, N.Y. 10018
or call: 212 302-6424
FAX: 718 961-1630

SPECIAL NOTICE

Tariffs, services and prices listed in this book are used for illustrative purposes only, and are subject to change. All tariffs, services and prices listed should be verified with the company in question, and/or the appropriate state public utility commission. Questions regarding interstate tariffs should be directed to the carrier in question, or to the Federal Communications Commission.

INTRODUCTION

Do you have to be an expert in billing to get refunds from the telephone company? Absolutely not! I know of a client who obtained a refund of $65,000 for her employer simply by knowing how to understand basic telephone company billing codes. This ability allowed her to identify billing data lines to a satellite location that had closed five years earlier.

The telephone company makes it difficult for its customers to understand their bills by the widespread use of jargon and technical codes. Once you know what these codes mean, you will be surprised at how easy it is to detect and correct errors that appear on your bill. This book explains the basics of telephone company billing procedures and shows you easy, non-technical steps that you can take to verify the accuracy of your bill. It will also show you ways to reduce your telephone bills. Basic services like Centrex, digital and analog private lines, DIDs and T1s are explained and detailed in easy to follow figures and charts. This book uses real life cases to show you how other companies have saved money and received refunds from their local and long distance carriers.

BACKGROUND

Prior to January 1, 1984 AT&T was responsible for the management of the Bell System. Throughout its reign, AT&T attempted to standardize the ordering and billing process for telephone service across the United States. AT&T's ultimate goal was to develop a seamless ordering and billing process whereby employees could transfer from state to state and still have the same systems and procedures at their new location.

AT&T came close to their goal. Over the years many procedures became common. By the late 1970s, the New York Telephone Company had been commissioned to develop a standard service order (SOP) and billing system (Customer Record Inquiry System or CRIS) for use throughout the Bell System. CRIS is the system that is used to bill business and residence customers, while CABS (the Carrier Access Billing System) bills the long distance carriers. The Beta company selected for initial deployment of the common billing system was the New England Telephone Company. As soon as Divestiture, the separation of the Bell System into seven regional companies, was announced in 1982, the project was called off. As luck would have it, the New York Telephone Company and New England Telephone Companies were reunited under the ownership of NYNEX, one of the seven Regional Bell Operating Companies created by the Divestiture Agreement. (For a complete listing of the companies created by the Divestiture Agreement consult the Glossary). New York and New England Telephone currently share both a common billing and service order system.

Even though a completely standardized system was never deployed, all the Bell companies continue to utilize billing and service order systems that are remarkably similar.

The Bell System set de facto standards for telecommunications for well over 100 years. Other telecommunications companies developed their own billing systems based on what customers were familiar with in their dealings with the Bell System. Even though much has changed since 1984, the basic format and design of telecommunications bills remain rooted in the Bell System. Even the new Competitive Local Carriers (i.e. Teleport and MFS) utilize many of the Bell formats and billing standards to make comparison of rates and services easier for their customers. In many ways, understanding the Bell System's billing logic is like learning Latin. It is the root of many offshoot languages, and thereby helps one better understand these offshoot languages. This book illustrates these concepts and has practical examples and examinations of actual bills from a variety of telecommunications companies.

The primary Regional Bell Operating Companies used as examples are Pacific Telesis, NYNEX and Bell Atlantic.

NYNEX is the Regional Bell Operating Company for the following local telephone companies:

1. New England Telephone Company

2. New York Telephone Company

New York Telephone's specific billing practices and procedures are used to detail how NYNEX bills for telephone service within its region.

Bell Atlantic is the Regional Bell Operating company for the following local telephone companies:

1. Bell of Pennsylvania
2. C & P Telephone Company of Maryland
3. C & P Telephone Company of West Virginia
4. Diamond State telephone Company
5. New Jersey Bell

New Jersey Bell's specific billing practices and procedures are used to detail how Bell Atlantic bills for telephone service within its region.

Pacific Telesis is the Regional Bell Operating company for the following local telephone companies:

1. Pacific Bell
2. Nevada Bell

Pacific Bell's specific billing practices and procedures are used to detail how Pacific Telesis bills for telephone service within its region.

Numerous other examples of billing practices and procedures from a variety of different telephone companies are included within this book.

CHAPTER ONE:

UNDERSTANDING YOUR LOCAL TELEPHONE BILL

Your Local Billing Area or LATA

Your local telephone company (telco) bills you for calls and services within a LATA. A LATA (Local Access Transport Area) is an area that shares a common economic or social purpose, and therefore is assumed to have a high degree of calling between residential and business telephone subscribers within that area. The Bell companies are prohibited by the Divestiture agreement from providing services that cross LATA boundaries. AT&T is prohibited from providing intraLATA services.

LATAs are population driven and often default to state and/or metropolitan boundaries. For example, California has eleven LATA's (see figure 1.1) while South Dakota has just one (LATA 640). Figure 1.2 lists all the LATA's that comprise the United States and its territories.

A LATA consists of a series of telephone numbers. For example, in California, LATA 724 consists of all the telephone numbers starting with an NPA (another word for area code) of 916 and with an NXX (the first three digits of your seven digit telephone number) listed in figure 1.3. Calls between these telephone numbers are considered intraLATA calls.

FIGURE 1.1

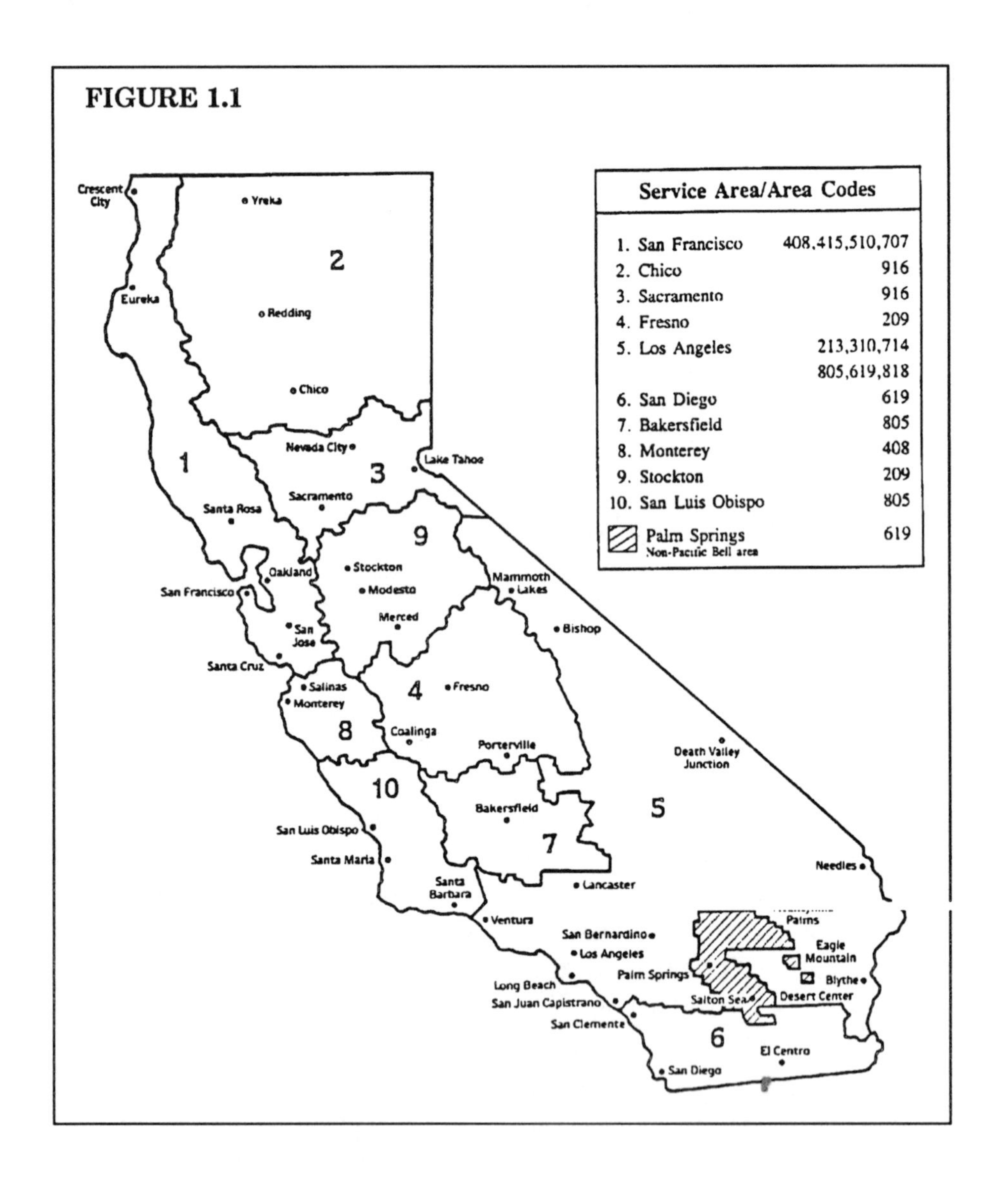

Your Telephone Number

To understand telephone terminology you must first understand the components of your telephone number. Your ten digit telephone number (for example 916 999-1234) consists of three separate components:

1	2	3
916	999	1234
NPA	NXX	LINE NUMBER

FIGURE 1.2

LATAs in the United States

LATA #	AREA	STATE	LATA #	AREA	STATE	LATA #	AREA	STATE
120	Maine	ME	430	Greenville	SC	635	Cedar Rapids	IA
122	New Hampshire	NH	432	Florence	SC	636	Fargo	ND
124	Vermont	VT	434	Columbia	SC	638	Bismark	ND
126	W. Massachusetts	MA	436	Charleston	SC	640	South Dakota	SD
128	E. Massachusetts	MA	438	Atlanta	GA	644	Omaha	NE
130	Rhode Island	RI	440	Savannah	GA	646	Grand Island	NE
132	New York Metro	NY	442	Augusta	GA	648	Great Falls	MT
133	Poughkeepsie	NY	444	Albany	GA	650	Billings	MT
134	Albany	NY	446	Macon	GA	652	Idaho	ID
136	Syracuse	NY	448	Pensacola	FL	654	Wyoming	WY
138	Binghamton	NY	450	Panama City	FL	656	Denver	CO
140	Buffalo	NY	452	Jacksonville	FL	658	Colorado Springs	CO
220	Atlantic Coastal	NJ	454	Gainesville	FL	660	Utah	UT
222	Delaware Valley	NJ	456	Daytona Beach	FL	664	New Mexico	NM
224	North Jersey	NJ	458	Orlando	FL	666	Phoenix	AZ
226	Capital	PA	460	Southeast	FL	668	Tucson	AZ
228	Philadelphia	PA	462	Louisville	KY	670	Eugene	OR
230	Altoona	PA	464	Owensboro	KY	672	Portland	OR
232	Northeast	PA	466	Winchester	KY	674	Seattle	WA
234	Pittsburgh	PA	468	Memphis	TN	676	Spokane	WA
236	Washington	DC	470	Nashville	TN	20	Reno	NV
238	Baltimore	MD	472	Chattanooga	TN	721	Pahrump	NV
240	Hagerstown	MD	474	Knoxville	TN	722	San Francisco	CA
242	Salisbury	MD	476	Birmingham	AL	724	Chico	CA
244	Roanoke	VA	478	Huntsville	AL	726	Sacramento	CA
246	Culpeper	VA	480	Montgomery	AL	728	Fresno	CA
248	Richmond	VA	482	Jackson	MS	730	Los Angeles	CA
250	Lynchburg	VA	484	Biloxi	MS	732	San Diego	CA
252	Norfolk	VA	486	Sheveport	LA	734	Bakersfield	CA
254	Charleston	WV	488	Lafayette	LA	736	Monterey	CA
256	Clarksburg	WV	490	New Orleans	LA	738	Stockton	CA
320	Cleveland	OH	492	Baton Rouge	LA	740	San Louis Obispo	CA
322	Youngstown	OH	520	St. Louis	MO	820	Puerto Rico	PR
324	Columbus	OH	521	Westphalia	MO	822	U.S Virgin	
325	Akron	OH	522	Springfield	MO		Islands	VI
326	Toledo	OH	524	Kansas City	MO	832	Alaska	AK
328	Dayton	OH	526	Fort Smith	AR	834	Hawaii	HI
330	Evansville	OH	528	Little Rock	AR	836	Midway/Wake	US
332	South Bend	IN	530	Pine Bluff	AR	920	Connecticut	CT
334	Auburn/		532	Wichita	KS	921	Fishers Island	NY
	Huntington	IN	534	Topeka	KS	922	Cincinnati	OH
336	Indianapolis	IN	536	Oklahoma City	OK	923	Mansfield	OH
338	Bloomington	IN	538	Tulsa	OK	924	Erie	PA
340	Detroit	MI	540	El Paso	TX	927	Harrisonburg	VA
342	Upper Peninsula	MI	542	Midland	TX	928	Charlottesville	VA
344	Saginaw	MI	544	Lubbock	TX	929	Edinburg	VA
346	Lansing	MI	546	Amarillo	TX	932	Bluefield	VA
348	Grand Rapids	WI	548	Wichita Falls	TX	937	Richmond	IN
350	Northeast	WI	550	Abilene	TX	938	Terre Haute	IN
352	Northwest	WI	552	Dallas	TX	939	Fort Myers	FL
354	Southwest	WI	554	Longview	TX	949	Fayetteville	NC
356	Southeast	WI	556	Waco	TX	951	Rocky Mount	NC
358	Chicago	IL	558	Austin	TX	952	Gulf Coast	FL
360	Rockford	IL	560	Houston	TX	953	Tallahassee	FL
362	Cairo	IL	562	Beaumont	TX	956	Bristol	TN
364	Sterling	IL	564	Corpus Christi	TX	958	Lincoln	TN
366	Forrest	IL	566	San Antonio	TX	960	Coeur D'Alene	ID
368	Peoria	IL	568	Brownsville	TX	961	San Angelo	TX
370	Champaign	IL	570	Hearne	TX	963	Kalispell	MT
374	Springfield	IL	620	Rochester	MN	973	Palm Springs	CA
376	Quincy	IL	624	Duluth	MN	974	Rochester	NY
420	Asheville	NC	626	St. Cloud	MN	976	Mattoon	IL
422	Charlotte	NC	628	Minneapolis	MN	977	Galesburg	IL
424	Greensboro	NC	630	Sioux City	IA	978	Olney	IL
426	Raleigh	NC	632	Des Moines	IA	980	Navajo	AZ
428	Wilmington	NC	634	Davenport	IA	981	Navajo	UT

FIGURE 1.3

LATA 724 **NPA/NXXs**

NPA-916

NXX	NXX	NXX	NXX	NXX
221	279	398	585	842
222	281	435	589	865
223	283	436	595	868
224	284	459	596	872
225	286	467	597	873
233	294	468	623	877
234	299	469	625	882
235	335	472	627	891
238	336	474	628	893
241	337	475	629	894
243	342	493	664	895
244	343	496	667	896
245	345	520	675	898
246	347	521	679	899
251	352	524	778	926
253	357	527	824	934
254	359	529	825	938
256	365	532	826	946
257	378	533	827	953
258	384	534	832	963
259	385	538	833	964
266	396	547	836	968
275	397	549	839	982
				993
				994

1. NPA or Area Code — Stands for the North American Numbering Plan Area. There are over 200 area codes in the United States, Canada, Bermuda, the Caribbean, and Northwestern Mexico. The middle number of the area code must be either a "1" or "0" (as in area code 212 or 203).

2. NXX — Your NXX identifies the specific telephone company central office building that provides service to a particular telephone number. NXXs are numbered as follows: N can be any number from 2 to 9 and X can be any number from 0 through 9.

NXX is often used interchangeably with the acronym CO (central office) or exchange. However, use of the term exchange usually has a broader meaning in that it refers to a specific switching center, i.e. the actual building housing a central office. This building serves a designated physical area with dial tone and may be sub-divided into a number of different central offices.

3. LINE NUMBER — The last four digits of your phone number which helps make it unique from all others.

Why Your Local Telephone Bill Is Frequently Wrong

When you order or disconnect a telephone line there are four basic steps to this process:

1. A telephone company representative enters an order into their Service Order Processor (SOP) system.

2. The order is then edited and passed to downstream provisioning systems (known by the telephone company as the FACS and TIRKS systems).

3. The actual physical work to connect or disconnect a telephone line is completed. The order is then passed to the Billing System - (CRIS).

4. Your bill is printed.

The telephone company representative must first take your order, and then manually assign a code(s) or USOC (Uniform Service Order Code) to your request. Each USOC corresponds to a particular telephone company service. The USOC code also tells the billing system at what price the service should be billed. This price should match that which the telephone company has tariffed (or advertised) the price at.

There are literally thousands of USOC codes and many are very similar in format but not necessarily similar in price. The more complex the order, the greater the chance for errors. While connecting or disconnecting telephone lines may be a four step process, some Bell companies have over eighty separate systems involved in the ordering, provisioning and billing process. Each one of these systems have their own internal edits or checks. Each system also has its own database of active telephone lines and circuits.

These systems often fall out of synch with each other and new service offerings may not be recognized in one or more of these systems. Obsolete codes may be purged from one system yet remain active in others. The telephone company often neglects end to end testing and most of its systems are tested separately.

These shortcuts often lead to errors. For example, if an order is rejected by one system it is held (or frozen) until the problem with the order is corrected. Many times a telephone company technician will disconnect a line (perform the physical work) even though the order is held by one of the software systems. This is done to avoid work delays. The feeling is that the customer is usually satisfied if the physical work is completed.

This method of action presumes that the order will be corrected at a later date (to reflect the work already done), but sometimes the volume of new orders becomes so heavy that the problem order is forgotten. In this case, the telephone bill never reflects the decrease in price for what was disconnected. It now becomes the responsibility of the customer to find and notify the telephone company of this error.

Special service orders (private lines, DIDs, T1s etc.) also contain many billing errors. Since these circuits usually connect two locations, it is possible for the billing for a circuit to continue even if one end of the circuit is physically disconnected. Chapter 2 explains the billing of special service orders.

Another reason why errors are so common has to do with human error and poor management at the telephone company. Most telephone company employees have enviable and relatively easy jobs. The telephone company representative that takes your order, however, has a difficult and demanding job. They are required by the telephone company management to "overlap" or perform several activities at once to increase productivity. For example, while they are talking to you on the phone, they may be filing paperwork unrelated to your telephone account. The representative's job is the most demanding and frustrating of all telephone company positions. As such it also has the highest employee turnover rate, which leads to further problems.

Another problem area involves rate changes. The telephone company is constantly applying for increases and/or changes to your monthly rates. When granted, they must be applied across all systems and "pro-rated" to reflect the proper start (or effective) date. Rate changes are often implemented at different times by different systems.

Each telephone company system has its own database. These databases are hopelessly out of synch. The bureaucracy at the telephone company often causes one department to be totally unaware of what a system managed by another department is planning to do. The result? Teleconnect Magazines states that over 90% of phone bills are wrong.

An entire industry has been built on auditing telephone bills and getting refunds for clients.

How To Decipher Your Local Bill

A typical local telephone company bill looks like the following:

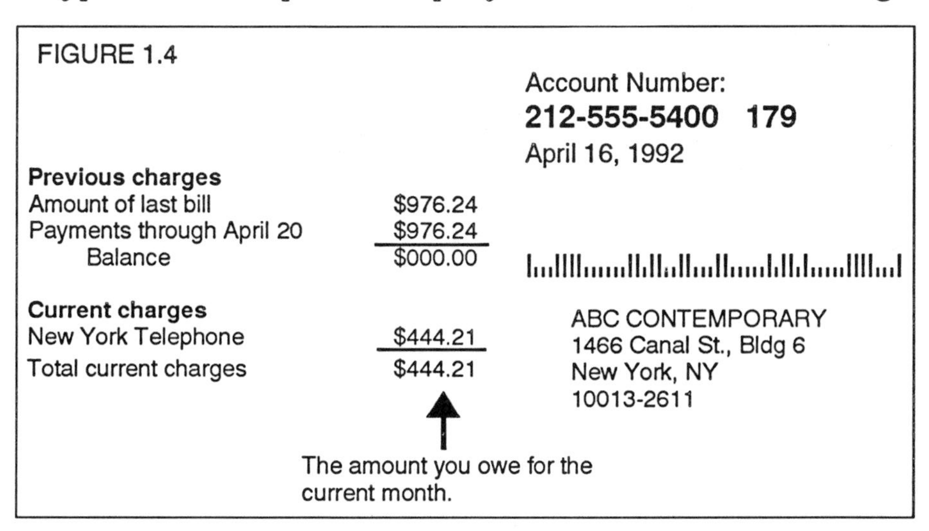

FIGURE 1.4

Account Number:
212-555-5400 179
April 16, 1992

Previous charges

Amount of last bill	$976.24
Payments through April 20	$976.24
Balance	$000.00

Current charges

New York Telephone	$444.21
Total current charges	$444.21

ABC CONTEMPORARY
1466 Canal St., Bldg 6
New York, NY
10013-2611

The amount you owe for the current month.

In this example, the customer owes the telephone company $444.21 and has a zero carry over balance from the previous month. How is that $444.21 derived? Let's analyze it step by step in order to understand the "basics". (Please refer to figure 1.5 on the following page).

1. BASIC SERVICE — Some telephone companies bill you in advance for the "rental" of the telephone line that runs between your business and their central office. The central office (CO) connects you to the telephone network.

FIGURE 1.5

New York Telephone
A NYNEX Company

Account Number 212 555 2400 179
April 16, 1992
Page 1

Helpful numbers
To question your New York Telephone charges call (212) 890-9999 regular business hours.
To order or change your service call (212) 890- 7777 regular business hours.
No charge applies on calls to the above telephone numbers if placed from within your Regional Calling Area or to an 800 number. Check the Customer Guide pages of your NYNEX telephone directory for more information.

Summary of New York Telephone charges

Basic service April 16 through May 15	$298.23	(1)
Service order/other charges and credits	5.28 CR	(2)
Directory advertising	30.50	(3)
Local calls	78.11	(4)
Directory information	2.43	(5)
County emergency services surcharge	2.10	(6)
Federal Tax (3%)	7.32	(6a)
State and Local taxes (8.25%)	30.80	(6b)
Total	$444.21	
		(7)

Basic service

These charges are for April 16 through May 15	$260.16	(8)
Line charge ordered by the Federal Communications commission	32.88	(9)
Wire maintenance charge (Optional)	5.19	(10)
Total	$298.23	

Other charges and credits

	Per month	Amount
Mar 25, 1992		
1. CR PRESUBSCRIPTION CHARGE		$5.00 CR#(11)
2. NY FCC SURCHARGE		.28 CR* (12)
	Other Charges and Credits Subtotal	5.28 CR
		Total $5.28 CR

Taxes: * Subject to Federal State/Local # Subject to State/Local

Directory advertising

	Per month	
Apr 01, 1992 through Apr 30, 1992		
1. MANHATTAN ALPHA	$30.50	$30.50 (13)
	Total	$30.50

Although the date of this bill is April 16th, the rental charge for service runs from April 16th through May 15. In this example, New York Telephone, a NYNEX company bills in advance. New England Telephone, also a NYNEX company, bills its customers in arrears.

2. SERVICE ORDER/OTHER CHARGES AND CREDITS — Since you are billed for basic service in advance by most telephone companies, you are owed a credit or refund if you should disconnect a line before the month is over. It is under this heading that the telephone company makes adjustments to your account. One time charges for installation or credits

for overbilling are also itemized under this heading.

3. DIRECTORY ADVERTISING — With your basic service, you typically get one listing in the white pages. If you want a bold heading or a larger advertisement, the charge for that advertisement is detailed here.

4. LOCAL CALLS — Usage is billed from the previous month. Calls made between March 16th and April 15th will be summarized under this heading. Local calls are those made within a LATA (intraLATA). Notice that unlike long distance bills you do not get an itemized detail of all your calls. Typically a charge of 50 cents per page applies to get an itemization of local calls. This itemization is called your Local Usage Detail (LUD) list.

5 DIRECTORY INFORMATION — Calls to the information operator (411 or 555-1212) are usually charged on a per call basis. In New York, these calls are billed at the rate of $.45 per call.

6. COUNTY EMERGENCY SERVICES SURCHARGE — This charge is used to help finance emergency services in the New York area. Most state legislators hide tax increases by embedding them within an already complicated telephone bill. Chapter 3 covers taxes and surcharges in more detail.

6A. FEDERAL TAX — This is a 3% federal excise tax levied by the federal government. Certain non profit companies are exempt from this tax.

6B. STATE AND LOCAL TAXES — In New York City state and local taxes are levied at a rate of 8 1/4 %. This tax varies according to your location. Certain non profit organizations are exempt from this tax.

7. BASIC SERVICE — This is your monthly rental charge for telephone lines and services. This book will teach you how to break down the basic service charge into its components and verify its accuracy. The overall basic service charge is divided into three main subheadings (8, 9, & 10).

8. BASIC SERVICE — $260.16 is the monthly charge you pay for use of the telephone company's facilities.

9. LINE CHARGE ORDER BY THE FEDERAL COMMUNICATIONS COMMISSION (FCC) — This charge was approved by the FCC to compensate the local telephone companies for loss of subsidies from long distance calling by the Divestiture agreement. In essence, it is an additional charge for you to be connected to the long distance network by the local telephone company. If you know what the monthly FCC Line Charge is in your state, you can determine the number of telephone lines you are billed for by dividing this number by the per line rate. For example, in New York the FCC Line Charge is $5.48 per line. Therefore this customer is paying a monthly charge for six telephone lines ($32.88 divided by $5.48). The FCC Line Charge is also known as the End User Line Charge or EUCL.

10. WIRE MAINTENANCE CHARGE (OPTIONAL) — Most telephone companies are required to inform you of services they provide that can be obtained elsewhere. You own the telephone wire (called inside wire) that lies within your premise. The telephone company provides an insurance policy whereby for a monthly charge they will repair a break in your wire at no additional charge. In this example the telephone company is identifying this service as above and beyond that which is required to make and receive calls.

I do not recommend this option, as inside wire rarely ever breaks (on average, once every 14 years). Most companies have service agreements with their equipment vendor that covers the replacing of inside wire. Before you agree to this service check with your equipment vendor.

Sometimes additional charges will mysteriously appear on your telephone bill. These additional services generate healthy profits for the local telephone companies. Recently, Robert Butterworth, Florida's Attorney General, charged Southern Bell with billing customers for inside wire plans that were never ordered or approved by their customers. These optional charges provide a significant revenue source to the telephone companies. If this charge suddenly appears on your bill you should immediately verify how it originated.

11. & 12. OTHER CHARGES AND CREDITS — This is an example of a credit this customer received for a presubscrip-

tion charge (charge to change long distance carriers). We will examine credits in more detail in Chapter 3.

13. DIRECTORY ADVERTISING — This is where the telephone company lists the actual directory that your advertising appears in.

We can quickly tell from this telephone bill that the customer's basic service consists of three separate charges of $260.16, $32.88 and $5.19, but how do we determine if this monthly basic service charge is correct? The first thing you need to do is request from the telephone company what is called a Customer Service Record (CSR).

The CSR breaks down your basic service charge into all its separate charges and components. By understanding how to read a CSR, you will be able to identify hidden and erroneous charges embedded within your basic service charge.

Your CSR is easily obtained by calling the telephone company's business office listed on your telephone bill. Simply ask for your CSR. The telephone company will not send out a CSR unless you specifically request it. The CSR is available free of charge and takes about 2 weeks to receive. No telcom manager should be without it.

One reason why the CSR is only mailed out on request is because few people (including most telephone company employees) can read it. It is an unsightly mix of computer codes and jargon. The sight of a CSR can be quite intimidating at first glance. Yet once you understand the billing codes on the CSR, they can actually be quite easy to read.

USOCs

The CSR is composed of USOCs which determine your fixed monthly rental charge. To fully understand a CSR, you need to enter the Bell System world of USOCs. A USOC (Uniform Service Order Code) identifies a particular service or equipment offered by the telephone company. When referring to it verbally, it is pronounced "U-Sock".

When you contact the telephone company for new service, a representative makes a record of your order. If, for example, you tell the representative that you want a business line with touchtone, he enters the USOC code of 1MB (Individual Message Line-Business) and TTB (touchtone business) into the SOP system. These codes tell the telephone company Provisioning and Plant departments to supply a business line with touchtone. These codes also tell the billing system to bill this customer business rates for the line and touchtone features as opposed to cheaper residence rates. (A new residence line would be entered as 1MR and TTR).

The reason why 1MB denotes individual message service business instead of the more logical code IMB is because the number 1 is always used instead of the letter I in USOC representation. Additionally, the letter O is used in place of the numeral 0. USOC codes are either three or five characters and can be alpha, numeric, or a combination of the two.

FIDs or field identifiers describe the attributes of a USOC much like an adjective describes a noun. For example, the FID PIC will always be somewhere to the right of the USOC 1MB. The FID PIC identifies who your long distance carrier is for a particular telephone line.

There are literally thousands of USOCs and FIDs in use throughout the Bell System. Though the Bell companies publish internal guides that define each individual USOC and FID (in English), some of the Bell companies will not send out copies of these guides to customers and consultants. To get a definition of a particular USOC you must call the telephone company's business office and specifically request a definition for each USOC in question. This is time consuming for both the telephone company and its customers.

The format of CSRs vary by Bell company. On some CSRs, USOCs are listed under columns labeled "ITEMS". FIDs sometimes appear under columns labeled "DESCRIPTIONS". When you first start reviewing CSRs ask the telephone company for help in identifying USOCs and FIDs on the CSR. As you gain experience you will easily identify them.

USOCs were introduced by AT&T in the 1970's to provide a platform for commonality between all the Bell telephone companies. As discussed earlier, AT&T wanted all its telephone companies to have a common service order processor and billing system. Since divestiture much has changed. The different and now independent Bell companies have introduced additional USOC codes for new services. Bellcore, which is jointly owned by the seven Regional Bell Operating Companies, is now responsible for encouraging and maintaining conformity among the Bell companies. Bellcore develops and approves the USOC codes used by these companies.

The trend is for Regional Bell Operating Companies like NYNEX and Ameritech to encourage the individual telephone companies that they manage to move toward strict commonality. There are cost savings in commonality between these telephone companies.

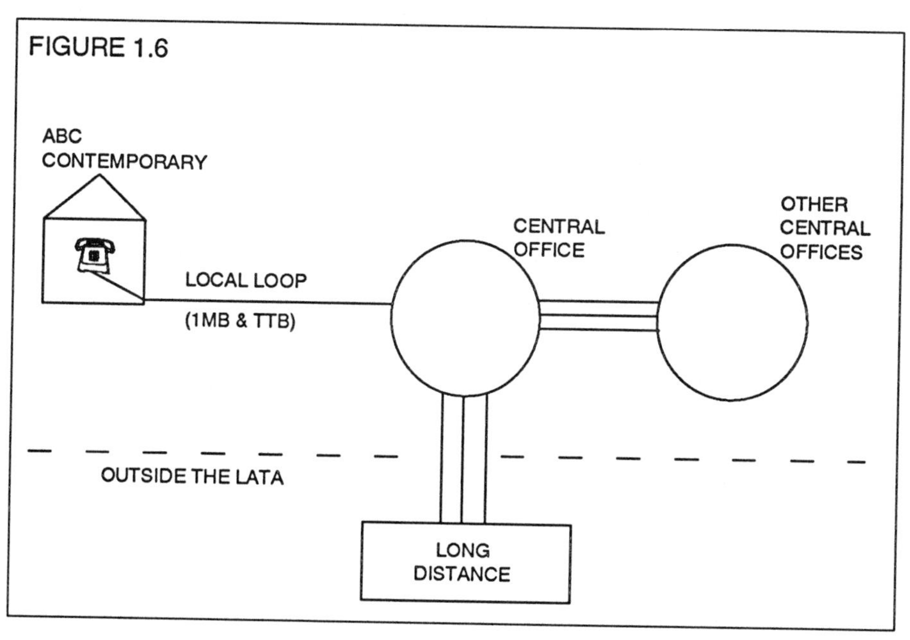

How USOCs Determine Monthly Billing Rates

Let's examine a schematic of a basic network configuration. (Please refer to figure 1.6 directly preceding).

A central office or CO is a telephone company switching facility where customers' lines are joined to switching equipment that connects customers directly to each other or to the long distance carriers.

The local loop connects your premise to the telephone company's central office. (In essence allows you to get dial tone).

You, in effect, rent this line from the telephone company. The USOC code for your local loop is (for a business customer) 1MB. 1MB stands for Measured Business Service. Simply stated , this code means that you will be billed a charge for each call you make, based on the distance and duration of the call. A flat rate would charge a set amount for a unlimited number of calls. This option, however, is often offered to residence customers, but is not usually available to business customers. New York Telephone will bill you $16.23 for the USOC 1MB.

The local loop (often used interchangeably with telephone line) can have features on it such as call waiting or touch tone. The USOC code for touchtone is TTB. (For illustrative purposes, we will continue to use New York Telephone's rates and codes. Later we will compare these codes and rates to other telephone companies). New York Telephone will bill you an additional $3.08 if you wish touchtone on your business line.

Each telephone line is also charged a FCC Line Charge for access to the long distance network. The USOC for the FCC line charge is 9ZR. New York Telephone charges $5.48 for the FCC Line Charge.

The total cost of one business telephone line from New York Telephone is:

USOC	MONTHLY CHG.
1MB	$16.23
TTB	$ 3.08
9ZR	$ 5.48
TOTAL	$24.79

The cost of a basic telephone line in New York is $24.79. Our example depicts a simple one line connection to the telephone network, commonly referred to as POTS service or "Plain Old Telephone Service". In our example, we took the charges for a telephone line and broke it down into its component parts. This process is known as unbundling.

When you have more than one telephone line at a location, the line that the bill is rendered under is referred to as your "mainline" or "main billing number", while each additional telephone line is referred to as an additional or auxiliary line.

Figure 1.7

CUSTOMER SERVICE RECORD

TN 212-555-5400

New York Telephone

TN (1)	CUS (2)	CD (3)	EX (4)	APP (5)	BD (6)	PRINT DATE	PAGE
212-555-5400	179		M		16	6/1/92	1

ORD (7)	CS (8)	SLS (9)	DD (10)	RECORD SEGMENT	PRINT REA
(E)	1MB			ACCOUNT	REQ

ACT	QUANT	ITEM (A)	DESCRIPTION (B)	L	ACTIVITY DATE (C)	RATE (D)	T	A
		LN (11)	A; B C CONTEMPORARY INC	I	3-31-89			
		LA (12)	1466 CANAL & 10013	#	10-24-83			
		SA (13)	1466 CANAL, MANHATTAN, NY+10013	I	12-20-90			
		LOC (14)	BLDG 6/DES ENT	#	10-24-83			
		YPH (15)	AK600 WOMENS APPAREL/CIS 5136	I	6/16/89			
			---BILL					
		BN1 (16)	ABC CONTEMPORARY		**_**_**			
		BA1 (17)	1466 CANAL ST BLDG 6		**_**_**			
		PO (18)	NEW YORK NY 10013		**_**_**			
		ADV						
		CCH	3					
		TAR (19)	002					
		EAR (20)	PRE RESTRUCTURE (2-1-90)					
			CHARGES					
			97.26					
			POST RESTRUCTURE (2-1-90)					
			CHARGES					
			119.43					
			---S&E (21)					
	1	BSX (22)	(CALLING CARD)	I	9-5-84		2	
	1	1MB (23)	/PIC MCI/PC CN,					
			09-16-91 ++FCC LINE					
		(24)	CHARGE++ (TIMED MESSAGE SERVICE)	T	7-28-86	16.23	1	
	1	RJ21X	/DES PU 5400,5402,9073					
			99-9249M 999-8928					
			(JACK-25 LN CONNECTOR)					
	1	TTB (25)	(TOUCH-TONE SERVICE)	I	6-8-87		4	
	1	ALN (26)	/TN 999-8928/PIC MCI	T	6-16-86	3.08	1	
			/PCA SN, 05-15-92 ++ FCC					
			LINE CHARGE ++ (ADDITIONAL LINE)	T	5-19-92	16.23	1	
	1	TTB	/TN 999-8928 (TOUCH-TONE SERVICE)	T	9-2-86	3.08	1	
	1	ALN	/TN 999-9249/PIC MCI					
			/PC SN, 05-15-92 ++FCC					
			LINE CHARGE++ (ADDITIONAL LINE)	T	5-19-92	16.23	1	
	1	TTB	/TN 999-9249 (TOUCH-TONE SERVICE)	T	4-7-86	3.08	1	
	1	ALN (27)	/TN 5401/PIC MCI/PCA CN,					
			09-16-91 ++FCC LINE					
			CHARGE++ (ADDITIONAL LINE)	T	7/5/85	16.23	1	
	1	ALN	/TN 5402/PIC/PCA CN,					
			09-16-91 ++ (ADDITIONAL LINE)	T	7-5-85	16.23	1	
	1	ALN	/TN 9073/PIC MCI/PCA SN,					
			03-16-92 ++FCC LINE					
		(28)	CHARGE++ (ADDITIONAL LINE)	T	3-18-92	16.23	1	
	1	MNTPB	/TN 9073 (BASIC WIRE MAINT- TYPE 1					
			WIRE)	T	4-1-91	5.19	4	
	1	TTB	/TN 9073 (TOUCH-TONE SERVICE)	T	2-24-87	3.08	1	
		HTG (29)	1 555.5400, 5401, 5402	I	2-24-87			
		(30)	---- CREDIT CARD					
		CHN	(A) JANE DOE	I	2-21-89			
		CHN	(B) LARRY JAMES	I	2-21-89			
		CHN	(c) PETE RICHARDS	I	2-21-89			
			---- IN SERVICE					
	6	9ZR (31)	FCC LINE CHARGE			32.88		

Lets examine a typical CSR (See Figure 1.7). The first two lines of a CSR correspond to information taken by the telephone company representative at the initial customer contact and entered into the SOP system. While these codes do not directly effect your bill, they are helpful to be aware of.

1. TN — Telephone number or main billing number

2. CUS — Stands for customer code. This is how the telephone company differentiates between customers who currently have a telephone number previously assigned to another customer.

3. CD — Completion date. The actual date your order completes or "posts" in the billing system.

4. EX — EX or exchange codes are used by the telephone to note if the customer accepts the appointment date offered by the representative or if he requests an expedite.

5. APP — The APP or appointment date is the date that the customer calls in to request service.

6. BD — Bill date. The date on which your bill will be calculated each month. In our example, the bill date is the 16th of every month.

7. ORD — Order number assigned to each service order request.

8. CS — Class of service. Here is the USOC 1MB that denotes this account as business service.

9. SLS — Sales code. The telephone Company uses this code to track which specific telephone company department initiated the service (for sales tracking purposes).

10. DD — Due date for the order to be installed. The customer may or may not accept the original installation date offered by the telephone company. The due date is the agreed upon installation date.

The next section of the CSR lists important information about your company. For example, it denotes how your company is listed with the Directory Assistance operator. If your company name is misspelled here, you could be missing out on quite a bit of business. This section is appropriately named the "LIST SECTION". Directly under the column labeled "ITEM" (column A) are USOCs, while the column labeled "DESCRIPTION" (column B) is made up of FIDs and contains more information about each USOC. The column labeled "QUANT" (column E) tells the quantity of USOCs billed.

11. LN — Listed name. Denotes how you are listed in the White Pages and in Directory Assistance.

12. LA — Listed address. Denotes the address detailed in the white pages and the address listed with the Directory Assistance operator.

13. SA — Service address. Sometimes the actual physical location (SA) of the telephone service is different from the listed address. The telephone company also tracks the actual location of service.

14. LOC — Location. Indicates the floor or level where the service is located.

15. YPH — Yellow page heading. Tells you under what heading your business is listed.

16. BN1 — Bill name. Lists who is responsible for bill payment.

17. BA1 — Bill Address. The address to which your bill is mailed.

18. PO — Post Office.

Over to the right of the CSR you will see a column (C) labeled "activity date". For each item (or USOC), a date is entered. This date is the last date activity occurred on this item. Each time a change is made to a USOC this date is updated and overlaid. If the USOC is never altered, then this date will be the original date the USOC was established. If a rate change occurs and the price of the USOC changes, then the activity date will reflect the date of this rate change.

19. TAR — Indicates the tax area you are in. This field is used by the telephone company to determine the state and local tax rate that is in effect in your area.

20. EAR — Special service rate restructure. Indicates the last date special service offerings were changed. Special service circuits are covered in chapter 2.

21. S&E — Service and equipment. This term is a throwback to the pre-divestiture days when the telephone company provided the telephone set along with your service. Today, most companies own or lease their equipment from vendors

other than their local telephone companies. In fact, the Bell Companies are prohibited from manufacturing telephone equipment under the terms of the Divestiture Agreement. At the beginning of the S&E section it is important to note that monthly rates (Column D) will appear to the right of most USOCs. This is where incorrect rates will show if a representative entered the wrong USOC when they took your order. Unless it is discovered and corrected by you, it will remain a perpetual monthly billing error.

As we will see, common errors include the partial disconnect of USOCs associated with a particular telephone line. For example, when you call to disconnect one of your telephone lines, the telephone company representative must remove all the associated USOCs (i.e. 1MB, TTB and 9ZR) to correctly reduce your bill by the proper amount.

22. BSX — Telephone calling credit card. In this instance the telephone company spells out, in English, what the USOC code translation is.

23. 1MB — We already know that 1MB is the charge for the local loop and its monthly rental fee is $16.23 (column D). In the description field the telephone line is described in more detail, but you need to have an even more detailed understanding of the telephone company's codes to grasp their implications on your service. To the right of the USOC 1MB is a line that reads /PIC MCI/PCA CN. Whenever you see a backward slash (/) the letters following this slash are referred to as a FID or field identifier.

FIDs always appear in the description field of the CSR. As previously noted, this field describes the USOC in more detail. /PIC indicates who you have presubscribed (chosen) as your long distance carrier (in this case MCI). Each telephone line has a separate entry for the long distance carrier. When you switch long distance carriers it is not uncommon for some lines to be missed. The customer often wonders why they are receive two (or three) different long distance bills. If that is the case, check each line to make sure all the PIC FIDs are the same.

We will not translate all the FIDs on our sample CSR, as

they do not directly effect the bill. As part of your audit, call the telephone company and obtain the English translation for all USOCs and FIDs that appear on the CSR. The knowledge you will gain will prove very helpful to you.

On the CSR, you will notice that the USOC 1MB does not identify the telephone number associated with it. This is because it is always the main billing number listed at the top of the CSR. In this example, the main billing number is 212 555-5400.

24. RJ21X — An RJ21X USOC denotes a standard telephone company jack that allows up to 25 telephone lines to terminate on it. This CSR lists the telephone numbers that terminate on this particular jack. No monthly charge is noted in column D for this USOC.

25. TTB — Touch tone business. The rate is $3.08 monthly.

26. ALN — Additional or auxiliary line. This customer has 999-8928 as one of his telephone numbers. Under the telephone company's numbering sequence, /TN stands for telephone number and both the NXX and line number are given only if the NXX differs from that of the main telephone number.

27. ALN — Here the telephone number is assumed to be 555-5401, as per the telephone company's numbering conventions.

Each 1MB and ALN basically repeats itself. A POTS line,though simple, still contains many opportunities for mistakes to be made. Some, as we shall see, are made in less than obvious ways.

28. MNTPB — Wire maintenance plan. Earlier we described this service as optional and that it is often a bad deal for the customer. As this customer only has a wire maintenance plan on line number 555-9073 it seems unusual and should be investigated. Call the telephone company and find out who authorized the charge. Ask yourself why would someone would place an order for a wire maintenance plan on only one line, and thus leave the other lines unprotected?

These are the kind of questions you should be asking yourself as you check CSRs. Look for what is out of the ordinary. Look for charges that seem to be double or in addition to the basic charges for a line.

29. HTG — Hunting or "rollover". Surprisingly, some local telephone companies do not charge for this feature, while others charge a substantial amount of money for this service. New York Telephone does not charge for hunting.

Hunting gives you the ability to have a call automatically transferred to another line after a certain number of rings without an answer or if a particular number is busy. This CSR tells us that when 555-5400 is busy, 555-5401 will ring and if 555-5401 is also busy, 555-5402 will ring.

The proper hunting sequence is critical to a business's operation and can only be verified by checking the CSR. Hunting is typically set up when telephone service is initiated, but is rarely updated as the business changes.

One of our clients told us that she had been receiving complaints from her customers that they would sometimes get the mail room when dialing her company's customer service number. She could not understand how this could happen sporadically. Upon examination of her CSR it was discovered that the telephone number for the mail room was listed at the end of her customer service hunt group. When all lines were simultaneously busy, overflow calls were being "rolled" to the mail room. You can imagine how much business was needlessly lost.

30. CREDIT CARD — Lists all the people in your company that have a credit card. If your firm has a high turnover rate, it is important to constantly review and update this information.

31. 9ZR — We have previously identified the USOC 9ZR as being the USOC for the FCC line charge. Each telephone line that you have is billed a 9ZR. In this example, the quantity column lists 6, so we can tell that the customer is billed for 6 telephone lines. The monthly rate of $32.88 can be divided by the quantity (6) to get the per line rate of $5.48.

What does this CSR tell us about our sample customer's telephone service? It tells us that the customer is being billed for six telephone numbers, 555-5400, 999-8928, 999-9249, 555-5401, 555-5402 and 555-9073 by the telephone company. Each line should have a touchtone charge. However, a review of the CSR reveals that telephone lines 555-5401 and 555-5402 do not have touchtone charges billed to them. This is what auditors look for. Note telephone lines that differ from the other telephone lines. Chapter 3 covers the methods you can use to find and correct common telephone company billing errors.

This CSR also tells us that the long distance carrier on each telephone line is MCI. In addition, line number 555-9073 is billed for a wire maintenance charge. We also know that if 555-5400 is busy, 555-5401 will ring, and then 555-5402 will ring if 555-5401 is busy (per the HTG USOC). Schematically, we can represent the service depicted by this CSR as follows:

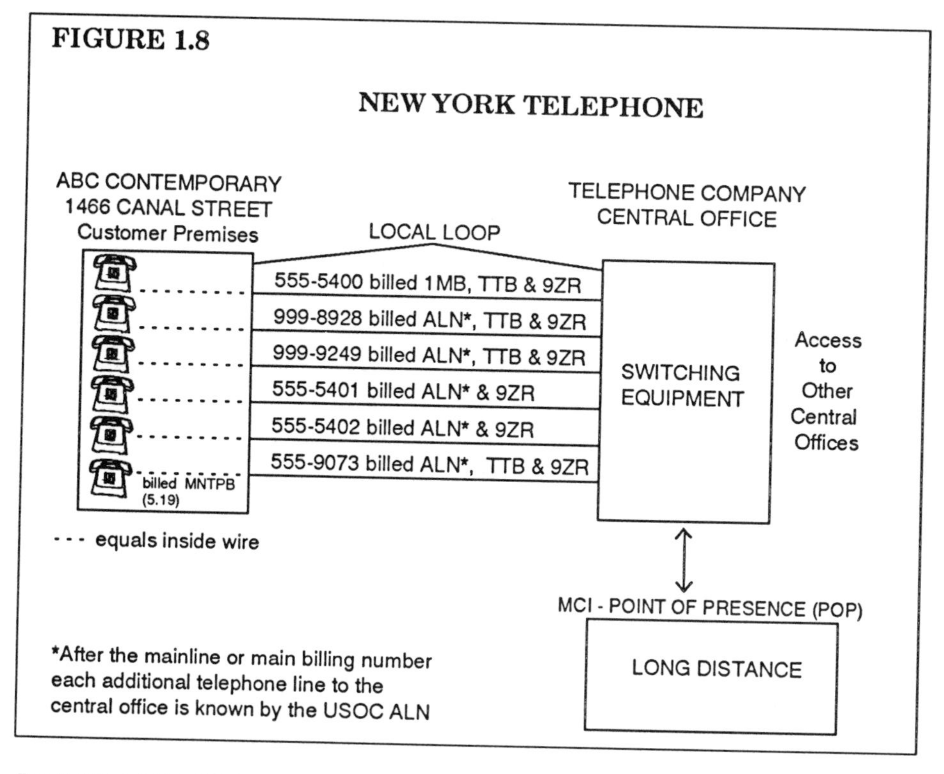

NEW JERSEY BELL

A CSR obtained from New Jersey Bell will differ from a New York Telephone CSR in that it is not printed on a clean, columned, lined paper. The Bell companies can be roughly divided into those companies that provide clean, easy to read CSRs versus those that print copies of what appears on their computer screens, staple them together and mail them out as CSRs.

I would wager that if their top executives were made aware of the condition of the CSRs mailed to customers, they would be shocked. CSRs that are local prints of computer screens do not portray an image of a high technology company. These CSRs often contain all sorts of extraneous information. To aid in your understanding of this type of CSR, we will ignore the extraneous information and zero in on what is pertinent to the charges on your bill. (See figure 1.9).

FIGURE 1.9

NEW JERSEY BELL CSR

201 555-5931 912 MAY 23, 1992 CSR P 1 13 1ML
GENERAL CONTRACTORS

Date	No.	Code	Description	Charge
07-20-91	(1)	LN	GENERAL CONTRACTORS INC	
07-20-91	(2)	LA	494 WILSON AV, NWK	
02-04-91	(3)	SA	496 WILSON AV, NWK	
	(4)	BN1	GENERAL CONTRACTORS	
	(5)	BA1	494 WILSON AV	
	(6)	PO	NEWARK NJ 07205	
			SERVICE AND EQUIPMENT	CHG
07-20-91	(7)	1ML	/PIC ATX/PICX 288/PCA NC, 02-01-91 /LPS/HML 0244/TER 1 / INDMESSAGE RATE BUS	12.84
07-20-91	(8)	HML	0244 - TER H1-13/DES 5931, 0602, 2711, 7949, 8147, 8296, 8386, 5045, 3123-3125, 3479, 3364/HTC /HML FID	NO CHG
02-04-91	(9)	TTB	/TOUCHTONE -PER LINE-BUS	1.99
07-10-91	(10)	9ZR	/FCC SUBSCRIBER LINE CHARGE	3.68
07-10-91			/DES TN 555-5932 POS 1, 555-3762	NO CHG
03-05-91	(11)	BSXUP	/CCSN 999 555	NO CHG
02-04-91	(12)	TTB	/TN 201 555-0602 /** TOUCHTONE PER LINE BUS	1.99
07-20-91	(13)	ALY	/TN 201 555-0602/LPS/PIC ATX/PICX 288/PCA NC, 02-04-91/HML 0244/TER 2/** AUXILIARY LINE	7.94
02-04-91		9ZR	/TN 201 555-0602 /** FCC SUBSCRIBER LINE CHARGE	3.68

As you can see, the New Jersey Bell CSR is not formatted very well, but the similarities in billing codes to New York Telephone are still evident.

1 to 6. Corresponds almost exactly to the LIST and BILL sections that appear in our sample New York Telephone CSR. See the New York CSR for explanations of these codes.

7. 1ML — 1ML is the equivalent to the 1MB USOC on the New York CSR. Notice, however, that the monthly rental cost is different. In New Jersey, the monthly rental for your main telephone line is $12.84. (The cost is located at the extreme right of the USOC 1ML). The customer has ATX (code for AT&T long distance as noted by the "/PIC ATX" FID).

8. HML — HML is the USOC for hunting in New Jersey. In this case the rollover sequence is: first 555-5931 rings, then 555-0602 then 555-2711 rings and so on.

9. TTB — The USOC denoting touchtone is the same in New Jersey as it is in New York. The monthly charge is $1.99 in New Jersey for touchtone.

10. 9ZR — Same USOC as New York, which denotes the FCC line charge, but at a different price.

11. CREDIT CARD — Has similar USOC to New York's, but New Jersey Bell lists the actual credit card number.

12. TTB — The next auxiliary telephone line also has touchtone.

13. ALY — The USOC code for an auxiliary or additional line is ALN. New Jersey Bell charges you more for the main billing line than for each additional auxiliary line. Over the years all of the Bell companies have developed their own billing oddities, over the years. Of course, there is no additional charge for New Jersey Bell to provide a main telephone line as opposed to providing an auxiliary line. The charge for each additional line is $7.94 while the main line is charged at $12.84 monthly.

COST COMPARISON BETWEEN NEW YORK TELEPHONE AND NEW JERSEY BELL FOR TWO POTS LINES:

NEW YORK TELEPHONE CO.

SERVICE	USOC	MONTHLY CHARGE
MAINLINE	1MB	$ 16.23
TOUCHTONE	TTB	$ 3.08
FCC LINE CHG.	9ZR	$ 5.48
AUXL LINE	ALN	$ 16.23
TOUCHTONE	TTB	$ 3.08
FCC LINE CHG.	9ZR	$ 5.48
TOTAL		$49.58

NEW JERSEY BELL

SERVICE	USOC	MONTHLY CHG.
MAINLINE	1ML	$12.84
TOUCHTONE	TTB	$ 1.99
FCC LINE CHG.	9ZR	$ 3.68
AUXL LINE	ALY	$ 7.94
TOUCHTONE	TTB	$ 1.99
FCC LINE CHG.	9ZR	$ 3.68
TOTAL		$32.12

PACIFIC BELL (CALIFORNIA) CSR

Pacific Bell's format of a CSR is also a local print from a terminal (See figure 1.10 on the following page). It is not as formal nor as neatly columned as our sample CSR from New York Telephone. Yet, it is not so cluttered as the CSR provided by New Jersey Bell. It is sort of a hybrid CSR. Pacific Bell is moving toward having an English language translation of each USOC listed directly on the CSR.

The BILL and LIST sections are easy to read and are also easily understandable. Pacific Bell first bundles services together then summarizes the billing for your services at a high level. On the end of the same CSR, they unbundle the billing for each billing component.

1. HTG — Unlike New York Telephone or New Jersey Bell, Pacific Bell charges for hunting at $.50 per line. The description field on the Pacific Bell CSR notes each line that has hunting.

2. 1ML — The charge for a business line from Pacific Bell is $8.35 per line ($25.05 divided by 3). 1ML is the same USOC code used by New Jersey Bell to bill your main business line.

3. EVC — USOC for extended busy call forwarding. This service automatically forwards incoming calls to another prearranged telephone number when your line is in use. An EVC USOC is used when the calls are forwarded to a telephone number outside your local central office. It is billed at a monthly rate of $7.95.

4. LW — Local white pages advertising.

FIGURE 1.10

213 555-1580 501 JULY 02 92 CSR P 1-1 1ML

MEASURED RATE BUSINESS SVC ACCOUNT NO 213 555 1580 501 S0

BILL DATE JULY 2, 1992
BILL NAME ABC CONTEMPORARY
MAILING 111 EAST 12TH STREET
ADDRESS LOS ANGELES

YOUR SERVICE IS LOCATED AT: 111 EAST 12TH STREET LOS ANGELES

	ITEM	QTY	DESCRIPTION	
(1)	HTG	3	HUNTING	1.50
(2)	1ML	3	MEASURED RATE BUSINESS SVC	25.05
(3)	EVC	1	EXTENDED BUSY CALL FORWARDING MONTHLY SERVICE	7.95
(4)	*LW		DIRECTORY ADVERTISING TAXES AND SURCHARGES	15.00
(5)	9ZR	3	ACCESS FOR INTERSTATE CALLING THESE AMOUNTS CAN BE FOUND IN THE MONTHLY CHARGES SECTION OF YOUR BILL	11.82
			PACIFIC BELL PRODUCTS AND SERVICES	
(6)	HTG	3	HUNTING	1.50
(7)	HTG	1	HUNTING GROUP 1580,7992,9036	
(7A)	CPR	1	RESTRICTED RECORDS INDIVIDUAL LINES 213 555-1580	
(8)	1ML	1	MEASURED RATE BUSINESS SVC YOUR LONG DISTANCE CARRIER IS MCI	8.35
(8A)	PIC	1	10222	
	9ZR	1	ACCESS FOR INTERSTATE CALLING	3.94

5. 9ZR — FCC line charge. Same USOC as used by New York Telephone and by New Jersey Bell.

The CSR issued by Pacific Bell now recaps your monthly charges in more detail. Each USOC is separated and explained.

6.& 7. HTG — Pacific Bell lists each line and the order in which hunting takes place. When 555-1580 is busy, your call hunts to 555-7992 and then to 555-9036.

7A. CPR — Indicates that your records are restricted from being shared with non-regulated Pacific Bell telephone company salespeople. Many of the Bell companies are required by state legislation to offer you this option.

8. 1ML — Measured Rate Service which means that you pay for each call that you make. The charge for your main business line is $8.35. Pacific Bell no longer charges extra for touchtone service. Touchtone is so common and widespread that Pacific Bell includes touchtone with the price for the actual telephone line. Look for all the Bell companies to eventually adopt this pricing format.

8A. PIC — Pacific Bell lists your long distance carrier for each telephone line. Here Pacific Bell lists the actual access code for your carrier.

COST COMPARISON NEW YORK TELEPHONE AND PACIFIC BELL FOR TWO POTS BUSINESS LINES:

NEW YORK TELEPHONE

SERVICE	USOC	MONTHLY CHG.
Mainline	1MB	$16.23
Touchtone	TTB	$3.08
FCC Line Charge	9ZR	$5.48
Additional Line	ALN	$16.23
Touchtone	TTB	$3.08
FCC Line	9ZR	$5.48
Hunting	HTG	NO CHG
TOTAL		$49.58

PACIFIC BELL

SERVICE	USOC	MONTHLY CHG.
Mainline	1ML	$8.35
Touchtone	N/A	NO CHG.
FCC Line Charge	9ZR	$3.94
Additional line	1ML	$8.35
Touchtone	N/A	NO CHG.
FCC Line Charge	9ZR	$ 3.94
Hunting	HTG	$1.00 *
TOTAL		$25.58

* Hunting is billed at .50 cents per line.

The following is an example of installation charges for two pots lines with Hunting — New York Telephone versus Pacific Bell:

PACIFIC BELL

$70.75 Installation fee per line (two lines)	$ 141.50
$30.00 Per hunt line	$ 60.00
TOTAL	$ 201.50

NEW YORK TELEPHONE

Service Charge	56.00
Premise Visit Charge	$ 19.00
Line activation charge - $71.75 per line	$143.50
TOTAL	$218.50

The following summarizes monthly charges for two POTS business lines from various Bell Telephone companies:

PENNSYLVANIA BELL

SERVICE	USOC	MONTHLY CHG.
Mainline	DTLBX	$11.15
Touchtone	TTB	$ 1.90
FCC Line Charge	9ZR	$ 3.72
Additional Line	DTLBX	$11.15
Touchtone	TTB	$ 1.90
FCC Line Charge	9ZR	$ 3.72
Hunting	HTG	NO CHG.
TOTAL		$33.54

BELL SOUTH

SERVICE	USOC	MONTHLY CHG.
Mainline	1FB	$28.00
Touchtone	TTB	$ 1.00
FCC Line Charge	9ZR	$ 6.00
Additional Line	1FB	$28.00
FCC Line Charge	9ZR	$ 6.00
Touchtone	TTB	$ 1.00
Hunting	HTG	$28.56
TOTAL		$98.56

NEW ENGLAND TELEPHONE

SERVICE	USOC	MONTHLY CHG.
Mainline	1FB	$33.30
Touchtone	N/A	NO CHG
FCC Line Charge	9ZR	$ 6.00
Additional Line	1LB	$33.30
Touchtone	N/A	NO CHG.
FCC Line Charge	9ZR	$ 6.00
Hunting	HTG	NO CHG.
TOTAL		$78.60

INDEPENDENTS

About 20% of the United States is served by non-Bell local telephone companies. These are companies that avoided being bought out by AT&T during its pre-Divestiture expansion period.

While independent, they do not operate in a vacuum. AT&T still cast a long shadow over their development and billing practices. Since many corporations have offices in many different states, they expect some sort of conformity from the independents. Many of their billing procedures and components are based on Bell standards. For example, it is common to bill a POTS line for the line to their CO (main or auxiliary line), touchtone, hunting and for the FCC line charge.

CONTEL

CONTEL is an independent local telephone company that operates in California. CONTEL, like Pacific Bell, does not charge extra for touchtone. CONTEL bills two POTS lines at a monthly rate of $78.80. CONTEL provides an English language explanation for each service they provide.

Mainline	$31.65
FCC Line Charge	$ 6.00
Additional Line	$31.65
FCC Line Charge	$ 6.00
Hunting	$ 3.50
TOTAL	$78.80

A CSR from CONTEL for four business lines is printed as follows:

BILLING NO. 999-5551

Service and equipment charges — CONTEL

QTY	DESCRIPTION	AMOUNT
3	Measured key telephone business access line	94.95
1	Measured business access line	31.65
3	Business wire care plus	2.70
4	Charges for network Access for interstate calling Your long distance carrier is MCI	24.00
2	Hunting arrangement	3.50
Total	**Contel charges from 10-25-92 thru 11-24-92**	**156.80**

SOUTHERN NEW ENGLAND TELEPHONE

Southern New England Telephone, another independent local telephone company, bills as follows for two lines:

QUANTITY TOTAL	DESCRIPTION	MONTHLY CHG.	
1	Select-a-call rate business trunks	$18.11	$18.11
1	Select-a-call rate business lines	$18.11	$18.11
2	Rollover features	$ 2.84	$5.68
2	Business subscriber Line charge	$ 5.15	$10.30
TOTAL			$52.20

Comparison between Bell and independent local telephone companies. Two POTS lines with hunting: (See figure 1.11).

Figure 1.11

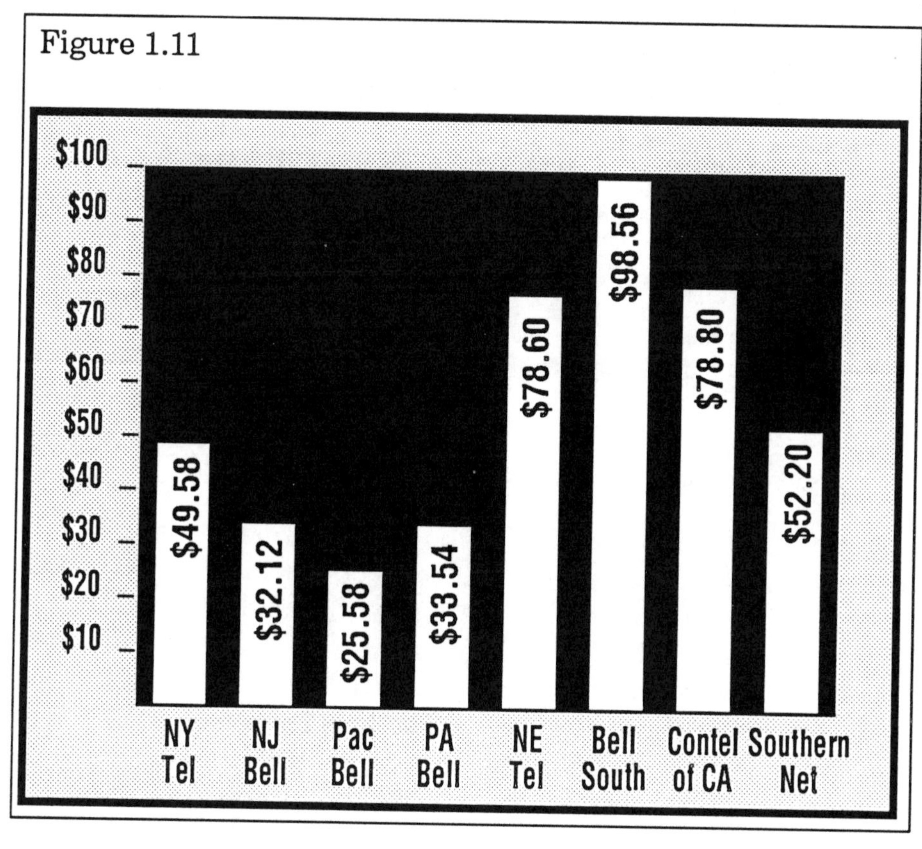

TRUNK VERSUS POTS LINE BILLING

Many telephone companies charge more for what is known as a trunk line. A trunk line is a central office line that is used in conjunction with a PBX, off premise extension or Direct

Inward Dialing (DID) service. The main difference between the two, is that PBX trunks are usually ground start while POTS lines are usually loop start. Ground start is where one side of a two wire trunk is momentarily grounded to get dial tone.

The only difference that justifies a higher cost for a ground start trunk is the application of the telephone company's Golden Rule #1— If you can get away with charging more for a particular service, charge more.

Some telephone companies also charge extra for touchtone on trunk lines as opposed to touchtone on POTS lines. In New York, for example, touchtone on a trunk line is $4.87 a month while only $3.08 on a POTS line.

When you are performing an audit you should check to see that trunk rates are not being applied to POTS lines. If a telephone line does not go through a switchboard then it should not be billed at a higher monthly rate. You can identify trunk lines because they are charged at a higher rate and will have distinct USOCs assigned to them. On a CSR from New York Telephone, trunk lines are billed as follows:

TXJ (USOC for trunk line)	$16.23
TJB (USOC for trunk touch tone)	$ 4.87
9ZR (FCC line charge)	$ 5.48
TOTAL	$26.58

CHAPTER SUMMARY

Your monthly charge for service from the telephone company consists of the charge for the local loop (1MB or ALN), touch tone (TTB) and the FCC line charge. You should now be in a position to read a CSR and identify all the telephone numbers that the telephone company is billing to your account. Chapter 3 will show you how to use this information to verify your billing, how to get the telephone company to correct your records and how you can obtain credit refunds from the telephone company.

CHAPTER TWO

SPECIAL SERVICE BILLING

INTRODUCTION

Special services can be broadly defined as any service provided by the telephone company that is not POTS. Examples of special service circuits include Direct Inward Dial (DID) trunks and T1s. Special service orders often require custom engineering and design work on the part of the telephone company.

A common special service offering is a dedicated point to point voice or data line, often referred to as a private line. It is called a private line because you have exclusive use of the line. This line bypasses the switched network. With a POTS line each time you call someone, a new connection is temporarily established to that location. The connection lasts as long as the call takes place. You are billed for usage based on the distance between locations and the time this connection stays in place.

A dedicated line also connects two locations. The connection, however, remains in place at all times. You pay a higher monthly fee for this type of connection (also based on the distance between locations, called mileage), but you are not billed a separate usage charge. You can make unlimited calls at a flat rate. It is often utilized by companies that have multiple offices within a certain geographic location. The more these offices call each other, the greater the justification for a private line. (See figure 2.1).

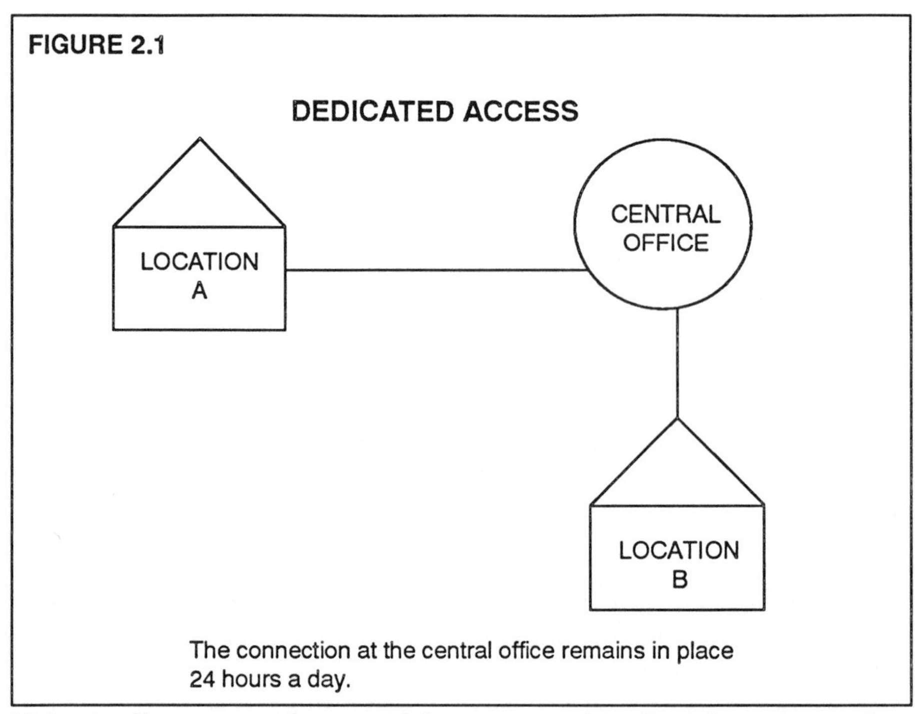

The connection at the central office remains in place 24 hours a day.

Private lines are also used by companies that have a mainframe computer at one location and multiple computer terminals at other locations. (In chapter 3 we will discuss how you can determine if a private line is cost justified).

The terms dedicated line, private line and point to point line are often used interchangeably. There are, however, many different types of data and voice lines that have various technological and pricing differences. This chapter will explain these differences.

Billing for special services usually appears at the end of the CSR, after billing for POTS lines are detailed. Some of the Bell companies bill circuits and private lines separately from your regular telephone bill on what is known as a special bill. Special bills often have a fictitious telephone number assigned to them. By billing special service circuits separately, attention is drawn to them and the customer is less likely to be billed for disconnected circuits.

Unfortunately, most of the local companies still include special service billing mixed in with your normal monthly telephone service. In these areas, you will find many errors. Unless you learn how to read a CSR you will never be able to determine if you are paying for disconnected private lines.

COMPONENTS OF SPECIAL SERVICE CIRCUITS

Special service circuits usually have four major components:

1. Circuit ID
2. Mileage
3. Loop Charges
4. Transmission & Signalling Requirements

1. CIRCUIT ID

Instead of a standard ten digit telephone number, special service circuits have an alpha-numeric means of identification. This identification is called the Circuit ID. All the Bell companies use a standard format to identify circuits. This naming convention is used throughout the industry and is often referred to as Bell System Common Language.

The Circuit ID can either be in telephone number format or in what is called serial number format. The /CLS FID after a circuit indicates the Circuit ID is in serial number format, while the /CLT FID identifies a Circuit ID that is in telephone number format. In our example below, the circuit ID is 96,OSNA,999240,,NY (serial number format).

The Circuit ID is divided into sections as follows:

	PREFIX	SERVICE CODE	MODIFIER	SERIAL NUMBER
POSITION #	1 2	3 4	5 6	7 8 9 10 11 12
CHARACTER	9 6	O S	N A	9 9 9 2 4 0

The number 96 (positions 1&2) and the serial number (positions 7 through 12) simply allows the telephone company to assign a unique number to each circuit. Positions 3 & 4 make up what is known as the service code. The service code identifies both the type of circuit and how the circuit is used. When taking your order for a special service circuit, the telco representative must assign a service code to your request. This allows the telephone company engineering department to design the circuit according to your needs.

Each special service circuit has unique needs and is charged accordingly. Therefore, it is critical for the telephone company representative to determine the proper code for your needs and requirements. The following chart lists some common service codes that you will find on a CSR.

SOME COMMON INTRA-LATA SPECIAL SERVICE CIRCUIT SERVICE CODES

BA	PROTECTIVE ALARM (DC)
BL	BELL & LIGHTS
CL	CENTREX CO LINE
CS	CHANNEL SERVICE (UNDER 300 BAUD)
CT	TIE TRUNK
CX	CENTREX CU STATION LINE
DA	DIGITAL DATA OFF-NET EXTENSION
DH	DIGITAL SERVICE
DI	DIRECT DIAL IN TRUNK
DJ	DIGITAL TRUNK
DK	DATA LINK
DO	DIRECT DIAL-OUT
DS	SUBRATE SPEEDS - DIGITAL DATA
DW	DIGITAL DATA - 56 KBS FACILITY
ES	EXTENSION SERVICE VOICE GRADE
FD	PRIVATE LINE-DATA
FR	FIRE DISPATCH
FT	FOREIGN EXCHANGE TRUNK LINE
FW	WIDEBAND CHANNEL
FV	VOICEGRADE FACILITY
FX	FOREIGN EXCHANGE LINE
HW	DIGIPATH II
LA	LOCAL AREA DATA CHANNEL

MT	WIRED MUSIC
ND	NETWORK DATA LINE
OC	CENTREX CU STATION LINE-OFF PREMISES
OI	OFF PREMISES INTERCOMMUNICATION LINE
ON	OFF NETWORK ACCESS LINE
OP	OFF PREMISES EXTENSION
OS	OFF PREMISES PBX STATION LINE
PA	PROTECTIVE ALARM (AC)
PL	PRIVATE LINE - VOICE
PM	PROTECTIVE MONITORING
PX	PBX STATION LINE
RA	REMOTE ATTENDANT
RT	RADIO LAND LINE
SG	REMOTE METERING-SIGNAL GRADE
TR	TURRET OR AUTOMATIC CALL DISTRIBUTOR (ACD) TRUNK
TL	NON TANDEM TIE TRUNK
WO	WATS LINE OUT
WS	WATS TRUNK OUT
WX	WATS TRUNK 2-WAY
WZ	WATS LINE 2-WAY
ZA	ALARM CIRCUITS

To find out what the service code “OS” signifies, look at the previous chart (chart 2.1). The service code OS is listed as an off premises PBX station line. This customer has a PBX switch and a dedicated connection (circuit) to a satellite location. The telephone at the satellite location operates as an extension off the PBX.

A tie line (a private line that connects two PBX switches) would be identified by the use of the service code TL. The Circuit ID would then read 96,TLNA,999240.

Position 5 of the circuit ID tells us the intended use of the circuit. Position 5 can either be the letter A, D, N or S. Each alpha character signifies something different. The following defines the use of each modifier.

A — Alternate data and non-data

D — Data

N — Non-data operation

S — Simultaneous data and non-data

In our original example Circuit ID (96,OSNA,999240,,NY), the 5th character is an "N" so we now know that the circuit is a off premise extension (service code OS), and is used for non-data (modifier N) purposes.

Character 6 indicates who provides certain facilities (the local telephone company, the long distance telephone company or your equipment vendor). Character 6 does not effect billing and is not covered in our discussion.

The N.Y. at the end of the Circuit ID tells us the circuit is operational in New York.

A Circuit ID, listed in telephone number format would appear as follows: 96,DINA,212,555,7400,001. Notice the similarity to a POTS telephone number. The format for a circuit in telephone number format is deciphered as follows:

	PREFIX	SERVICE	MODIFIER	NPA	CO	LINE #	EXTENSION
POSITION #	1 2	3 4	5 6	7 8 9	10 11 12	13 14 15 16	17 18 19
CHARACTER	9 6	D I	N A	2 1 2	5 5 5	7 4 0 0	0 0 1

2. MILEAGE

You pay more for a circuit that is located across town as opposed to a circuit that connects locations on different floors of the same building. Distance sensitive charges are covered under mileage billing. Mileage billing contains two major components. The first component is a fixed minimum charge that is always billed regardless of the distance between locations.

The second component is known as interoffice mileage (INOF in telephone jargon) charges, and is strictly distance based. You are actually billed for the distance (in air miles) between the telephone company CO that serves your location and the telephone company CO that provides service to the terminating (sometimes called the foreign location) end of the circuit. For example, New York Telephone bills mileage on circuits that travel 12 miles or less as follows:

	FIXED RATE	VARIABLE RATE
0 TO 12 MILES	$36.86	$ 3.17 PER 1/4 MILE

For circuits that travel in excess of 12 miles:

	FIXED RATE	VARIABLE RATE
OVER 12 MILES TOTAL	$36.86	$12.67 PER MILE

For example, a circuit that connects locations with COs that are 14 air miles apart will be charged as follows:

Fixed Mileage Charge	$ 36.86
Variable Mileage Charge 14 Miles At $12.67 Per Mile	$177.38
Total	$214.24

A circuit that travels 2 1/2 miles will be billed as follows:

Fixed Mileage Charge	$ 36.86
Variable Mileage Charge	
(10) 1/4 Miles At $3.17 per 1/4 Mile	$ 31.70
Total	$ 68.56

As the chapter progresses, we will encounter additional examples of mileage billing.

3. LOOP CHARGE

The loop charge covers the cost of the connection between each location and its local CO. A circuit will have two loop charges, one for each end. It is analogous to the (1MB) local loop charge described in chapter 1, except that a premium rate is charged for dedicated line service.

4. TRANSMISSION & SIGNALING

Transmission packages are based on the type of equipment that is connected to the circuit. Different design parameters are required from the telephone company depending on how and for what purpose a circuit will be utilized. Your equipment vendor specifies what type of transmission is required.

Signaling determines how one location of a circuit tells the other location that they are calling or sending data. It determines, for example, if a telephone will ring automatically at one location if someone picks up a telephone at the other location. New York Telephone bills for transmission packages under the heading of Feature Functions as follows:

Feature Function	**Monthly Rate**
Basic 2-Wire Voice	$ 7.07
Basic 4-Wire Voice	$19.99
Type A PXOS Feature	$12.77
Type B PXOS Feature	$12.77
Type C PXOS Feature	$ 6.57
Basic 2-Wire Data	$ 4.67
Basic 4-Wire Data	$11.65
Enhanced 4-Wire Data	
Type C-1	$11.65
Type C-2	$11.65
Type C-4	$11.65
Type D-1	$11.65

Analog Versus Digital Circuits

An analog circuit is often called a voice grade circuit because it operates in the frequency range that best carries voice conversations (300 to 3000 Hertz). An analog circuit can transmit data traffic by means of a modem. A modem, for example, will convert data from your PC (in zeros and ones) to a sine wave capable of being transmitted over analog lines. Digital circuits can transmit data without the use of a modem. Conversely, voice conversations must then be digitized to be sent over digital lines or circuits.

SPECIAL SERVICE CSRs

Armed with a basic knowledge of USOCs and the major components of a special service circuit, we can now analyze a CSR that contains billing for a special service circuit (see figure 2.2). The following explains the special service portion of a New York Telephone CSR. (Portions of the CSR are abbreviated).

FIGURE 2.2

SPECIAL SERVICE CSR

	QUANT	ITEM	DESCRIPTION	RATE
(1)	1	PFSAS	/CLS 96,OSNA,999240,,NY	
(2)	11	3LN2Y	/CLS 96,OSNA,999240,,NY	71.73
(3)		CKL	1-1466 CANAL ST., MANHATTAN /CLS 96,OSNA,999240 /LSO 212 555/TAR 002	
(4)	1	SN	ABC CONTEMPORARY	
(5)	1	CON2X	LOOP CHARGE 2-WIRE	21.53
(6)	1	PMWGX	TYPE C PXOS FEATURE	6.57
(7)		CKL	2-1466 BROADWAY, MANHATTAN 96,OSNA,999240,,NY /LSO 212 999/TAR 002	
(8)		SN	ABC CONTEMPORARY	
(9)	1	CON2X	LOOP CHARGE 2-WIRE	21.53
(10)	1	PMWGX	TYPE C PXOS FEATURE	6.57

1. PFSAS — This USOC identifies the circuit ID as 96,OSNA,999240,,NY. We can tell it is in serial number format from the /CLS FID. This customer has an private line that connects a telephone at another branch office to his PBX at their main location. This information is known by looking up the service codes OS on page. The service code OS is the code for a off premise PBX station line.

2. 3LN2Y — Prefixes like 3LN, 1LN or 1LB always indicate mileage billing. When the second character of the USOC is an "L", you can be sure that the USOC generates mileage billing. The fixed minimum mileage charge for this circuit is $36.86, and the interoffice mileage is billed in 1/4 mile increments at $3.17 per 1/4 mile.

Every telephone company bills for mileage at different rates. These rates can be obtained by checking the telephone company tariffs, or by calling the telephone company directly.

The number eleven (11) in the quantity field indicates that the distance between the central office serving the originating location and the central office serving the terminating location is 11 (1/4) miles or 2 3/4 miles. The mileage cost between central offices is determined by the distance in a straight line (actually calculated via V&H coordinates). Telephone company personnel will often use the term "as the crow flies" to describe the calculation of mileage between COs.

3. CKL — USOC denoting the circuit location of the (1) or originating location. In this example CKL 1 is 1466 Canal Street. The /LSO 212 555 indicates that the telephone company Local Serving Office or LSO (yet another word for central office) is 212 555. All telephone numbers that begin with 212 555 receive dial tone from that particular central office.

4. SN — This USOC denotes the listed name of the customer that is located at the CKL 1 location.

5. CON2X — USOC that denotes a 2-Wire loop charge. A 2-Wire circuit allows transmission to take place in one direction at a time. A 4-Wire circuit permits simultaneous transmission between locations.

6. PMWGX — This USOC denotes the type of signaling on the circuit, i.e. automatic ringing of the telephone at the terminating end when it is activated.

7. CKL 2 — Refers to the distant CKL 2 or terminating location of the circuit. This circuit terminates at 1466 Broadway, and its LSO is 212 999.

8. SN — USOC that denotes the listed name of the customer at the terminating end. In this example the same company has offices at both locations.

9. CON2X. — Repeat of USOC listed in number 5.

10. PMWX — USOC that indicates the type of signalling used by the circuit.

Schematically this circuit is depicted as follows (See figure 2.3 on the following page):

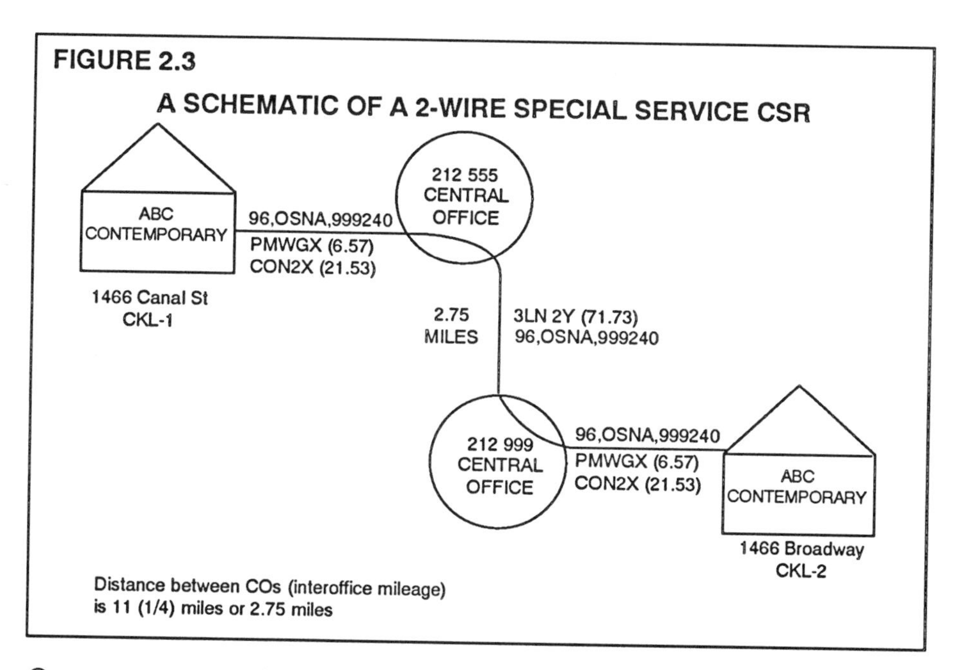

Our next example (see chart 2.4), depicts a CSR that bills a 4-Wire, full duplex data line. The USOC CON4X is the USOC associated with a 4-Wire circuit. Notice that a 4-Wire circuit bills a higher monthly charge than a 2-Wire circuit.

CHART 2.4

A CSR THAT BILLS A 4-WIRE CIRCUIT

QUANT	ITEM	DESCRIPTION	RATE
		SPECIAL SERVICE	
1	PFSFS	/CLS 74,FDDA,99971,,NY (PRIVATE LINE FACILITIES)	
14	1LNAS	/CLS 74,FDDA,99971,,NY (MILEAGE CHARGE)	214.24
	CKL	1-179-25 ROCKAWAY BLVD. /CLS 74,FDDA,99971,,NY /LSO 718 555/TAR 600	
1	CON4X	/EAR (LOOP CHARGE 4-WIRE)	40.61
1	PMWD4	/EAR (BASIC 4-WIRE DATA CIRCUIT)	11.65
	CKL	2-89 VEASEY ST. NY NY /CLS 74,FDDA,99971,,NY /LSO 212 999/TAR 002	
1	CON4X	/EAR (LOOP CHARGE 4-WIRE)	40.61
1	PMWD4	/EAR (BASIC 4-WIRE DATA CIRCUIT)	11.65

The first step in an audit of this CSR, would be to verify that this circuit is actually a 4-Wire circuit. The required design of a 4-Wire circuit can usually be verified with the equipment vendor that initially ordered it.

To better understand special service CSRs, take a minute to analyze the Circuit ID listed in Chart 2.4. Start by looking at the service code (position 3&4) of the Circuit ID. In this example the Circuit ID is 74,FDDA,99971,,NY, where FD is the service code. Now, look back at the service code list (page 38&39) and look up "FD". "FD" is the service code that denotes private line-data. The "D" in position 5 indicates its use is for data only. This type of private line is often used to transmit data from a computer at one location to a computer at another. With this knowledge in hand, you have a better chance of finding the department who originally ordered the circuit. You can then determine if the circuit is still in use.

You may also notice that the mileage billing for this customer is different. For circuits in which the originating and terminating end are 12 miles or greater apart, mileage is billed in increments of full miles versus 1/4 miles. In this case the quantity is 14 (for 14 miles). The USOC 1LN4S directs the billing system to bill mileage at a different rate. The fixed minimum charge is still $36.86, but the per mile rate is $12.67. The total mileage cost is determined by multiplying the number of miles (14) times $12.67, and then by adding this billing for interoffice (INOF) mileage cost to the fixed minimum mileage cost.

Schematically our customer is detailed as follows (See figure 2.5):

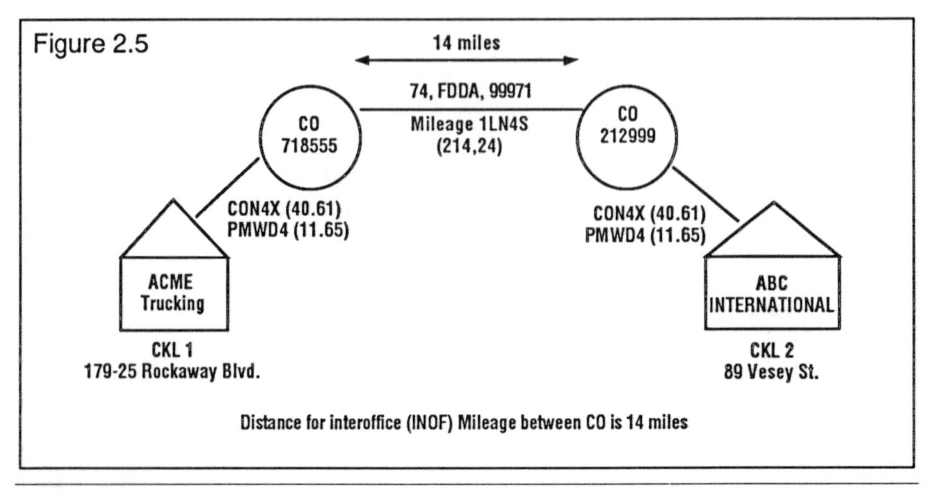

TIE LINE — PBX TO PBX 4-WIRE CONNECTION;

The following summarizes the monthly and installation charge by New York Telephone for a 4-Wire tie line connecting 2 PBXs. In this example, the originating and terminating COs are 1/2 mile apart.

Installation for this circuit (using New York Telephone rates) will cost you a one time fee of $1,272.64. Your monthly cost will be $200.80. This fee allows you to transfer an unlimited number of calls between locations at no additional cost.

	MONTHLY	INSTALLATION
SERVICE ORDER CHARGE		$56.00
PREMISE VISIT CHARGES (ONE PER LOCATION)		$38.00
CO LINE CONNECTION CHARGES		$143.50
NRCCO (LOOP CONNECTION CHARGE		$318.00
CHANNEL CONNECTION		$473.32
CON4X (4-WIRE)	$81.22	$81.22
BASIC (4-WIRE) VOICE	$39.98	$39.98
SIGNALING ARRANGEMENTS	$36.40	$122.62
INOF MILEAGE		
FIXED	$36.86	
VARIABLE(1/2 MILE)	$6.34	
TOTALS	$200.80	$1,272.64

Schematically, this circuit is depicted as follows:

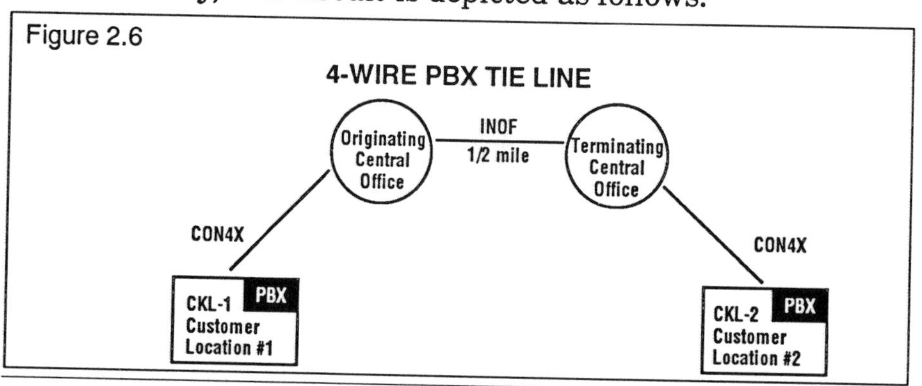

SPECIAL CIRCUIT BILLING VARIATIONS

Some Bell companies identify the originating and terminating (CKL 1 and CKL 2) locations of a special service circuit at the beginning of the CSR, within the LIST section. The CKL locations will appear after the LOC FID.

Each CKL location is usually assigned a letter of the alphabet. When the Circuit ID is listed in the S&E section, the CKL FID will have two alphas after it. You must refer back to the LIST section of the CSR to determine the originating and terminating locations of the circuit. The first alpha corresponds to the originating location and the second alpha denotes the terminating end. New Jersey Bell is a Bell company that uses this method of detailing circuits on its CSRs. The following example illustrates how the locations of the circuit can be identified:

FIGURE 2.7

DEPICTS A NEW JERSEY BELL CSR

201 555-5200

	05-24-91	LN	ABC CORPORATION	
	05-24-91	LA	1 PORT JERSEY BLVD, J CY	
	05-24-91	SA	302 PORT JERSEY BLVD, J CY	
	05-24-91	LOC	ENTIRE BLDG	
	06-26-86	CKL	(A) 27 PORT JERSEY BLVD J CY ;GUARDHOUSE	
	02-15-85	CKL	(B) CARGO AREA, PORT JERSEY BLVD, J CTY	
	02-15-85	CKL	(C) GUARDHOUSE #2, PORT JERSEY BLVD, J CY	
	02-15-85	CKL	(D) OFFICE, PORT JERSEY BLVD, J CTY	
	02-15-85	CKL	(E) ANNEX , PORT JERSEY BLVD, J CY	
	11-21-88	CKL	(P) 17 TERRACE DRIVE, UP SADL RI	
			SERVICE AND EQUIPMENT	CHARGES
	11-21-88	CKT	OSNA 999293NJ	
(1)		ARR	PX 4XAYS/CKL A, P	
			PVLLS /**PVT LN-VOICE GRADE-CPE STS EQP	NO CHG
(2)			9B1SX /CKL A/LSO 201 555/DES BLDG GUARDHOUSE/ LOC CHANNEL DIF EXCH-2014	23.46
(3)			9B1SX /CKL P/LSO 201 999/LOC CHANNEL DIF EXCH-2014	23.46
(4)		25	1LVA4 /SEC /EX J CY,RAMSY/** MILEAGE	92.07

When you review a New Jersey Bell CSR, you will notice slight Circuit ID naming variations. In the previous example OSNA 999293NJ is the Circuit ID. The Circuit ID is not prefixed with a number (i.e. 96 as with New York Telephone), nor do commas separate the components of the Circuit ID.

1. ARR — Lists the originating and terminating locations for Circuit ID OSNA 999293NJ. You must refer back to the LIST section to ascertain the locations of A and P. CKL (A) is located at 27 Port Jersey Blvd., in the Guardhouse, while CKL (P) is at 17 Terrace Drive, Upper Saddle River.

2 & 3. Loop charges — As with New York Telephone, New Jersey Bell charges a premium on the portion of your circuit that runs between the CO and each location.

4. Mileage component — As with New York Telephone, New Jersey Bell charges you for mileage based on the distance between the COs serving each end of the circuit. New Jersey Bell bills mileage at different rates based on the speed and design of the special service circuit. In this example, New Jersey Bell bills mileage at a rate of $11.67 for the first mile, and $3.35 for each additional mile (see Chapter 3 for more information).

Schematically our circuit looks like the following:

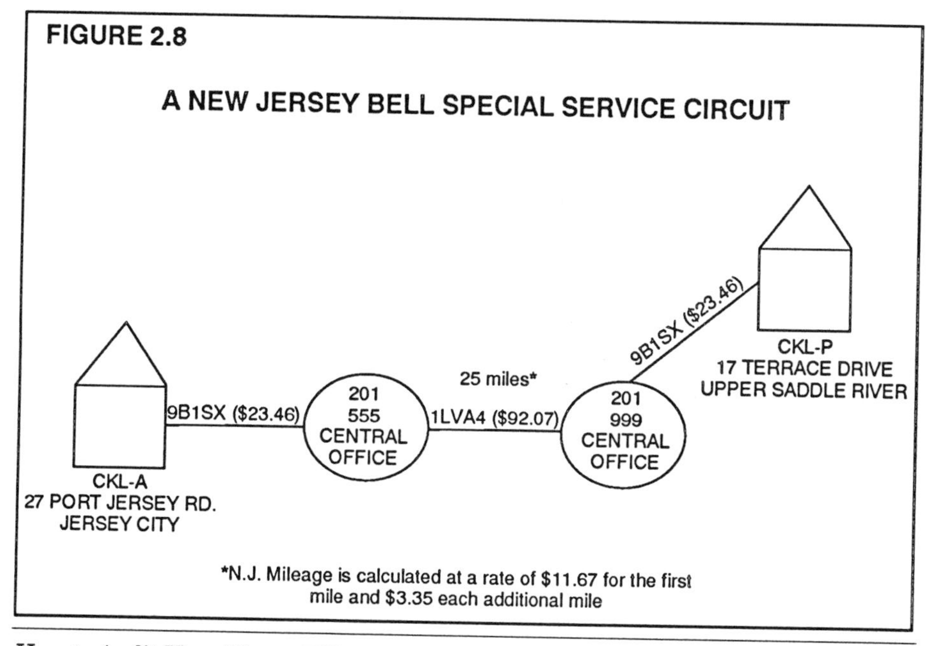

DIGIPATH DIGITAL SERVICE II

DIGIPATH, as offered by New York Telephone, is a digital circuit that can transmit data at speeds of 2.4, 4.8, 9.6, 19.2, and 56Kbps between multiple terminals and/or hosts.

The monthly billing and installation charges are fairly straight forward. A CSR (condensed in figure 2.9 for illustrative purposes) for a DIGITPATH circuit is depicted by New York Telephone as follows:

FIGURE 2.9

DIGITPATH CIRCUIT

QUANT	ITEM	DESCRIPTION	RATE
1	DYDJS	/CLS 96,HWDA,422999,,NY	
1	OCH	/CLS 96,HWDA,422999,,NY INTEROFFICE CHANNEL	$30.51
1	1LNPX	/CLS 96,HWDA,422999,,NY INTEROFFICE CHANNEL MILEAGE PER MILE	$4.91
	CKL	1-250 BROADWAY, MANHATTAN NY	
	SN	BROKERLAND INC.	
1	1DCPX	CHANNEL TERMINATION	$74.89
1	NRASQ	CIRCUIT TERMINATION CHARGE	
	CKL	2-2 WORLD TRADE CENTER, MANHATTAN, NY	
	SN	TELSTAR RESOURCES	
1	1DCPX	CHANNEL TERMINATION	$74.89
1	NRASQ	CIRCUIT TERMINATION CHARGE	

The billing for a DIGIPATH II circuit is summarized in figure 2.10. Figure 2.11 illustrates this circuit schematically.

FIGURE 2.10

DIGIPATH II CIRCUIT

QUANT	ITEM	MONTHLY	INSTALLATION
1	SERVICE ORDER		$56.00
2	PREMISE VISIT		$38.00
2	CHANNEL TERMINATIONS @ 74.89 EACH	$149.78	$1,054.82
	INTEROFFICE CHANNEL		
	FIXED	$30.53	
	VARIABLE	$4.91	
	TOTAL	$185.22	$1,148.82

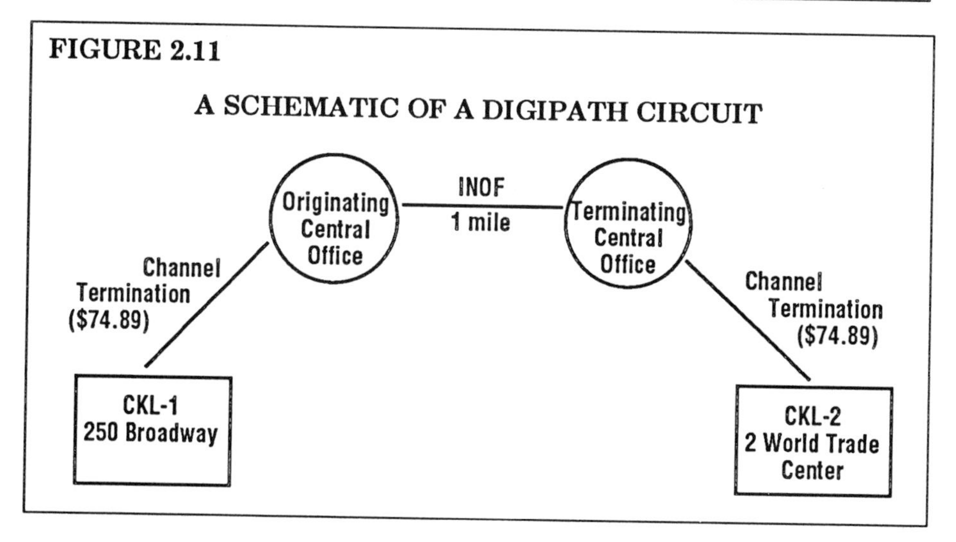

DID Trunks

Direct Inward Dialing (DID) trunks gives an outside caller direct access to a specific individual telephone extension that is connected to a PBX. The caller can directly dial a PBX extension without the assistance of an attendant. Direct Inward Dialing trunks also allow many different telephone numbers to share a single trunk (though only one telephone conversation can take place at any one time).

Customers are assigned Direct Inward Dialing telephone numbers in blocks of 20 or 100 numbers, i.e. 555-1000 to 555-

1020. DIDs improve efficiency by reducing the workload and holding times for telephone attendants. They also optimize PBX port usage since many telephone numbers are concentrated onto a limited number of trunks.

The components of a DID are similar throughout the Bell companies. The total cost of a DID trunk in New York is $80.73. Unlike a circuit that connects two locations, a DID trunk connects you directly to the CO.

You are charged a premium for the ability of a single DID trunk to recognize many different telephone numbers. Notice that a DID is charged for a 2-Wire loop charge as opposed to a 4-Wire loop charge. A DID trunk can only be used to receive incoming calls; you can not make outgoing calls over them. Hence, the 2-Wire (allows one way transmission) loop charge.

The following details how New York Telephone bills a DID trunk:

	USOC	RATE
DID Station Numbers	ND4	$3.64 PER 20 NUMBERS
DID Station Numbers	NDZ	$18.23 EACH ADDITIONAL GROUP OF 100#S
DID Trunk Charge	TB2	$56.17
DID Loop (2W)	D1F2X	$19.08
FCC Line Charge	9ZR	$5.48
TOTAL		$80.73

The following is a sample CSR for five DIDs

Figure 2.12

	QUANT	ITEM	DESCRIPTION	RATE
(1)	1	ZZYEB	/REF A/ZDID 212 555-1840 - 1899 (TOLL FILE - DID)	
(2)	20	ND4	/REF 1 /TA 12 MO, 06-20-92 /SBN 555-1840-1859 (DIRECT IN STA NUMBER)	
(3)	20	ND4	/REF 2 /TA 12 MO, 06-20-92 /SBN 555-1860-1879 (DIRECT IN STA NUMBER)	
(4)	20	ND4	/REF 3 /TA 12 MO, 06-20-92 /SBN 555-1880-1889	

			(DIRECT IN STA NUMBER)	
(5)	1	TB2	/TER 212 555-1840: 001	
			/TA 12 MO, 06-20-90	
			/LSO 212 555	
			/CLT	
(6)			96,DINA,212,555,1840,001	
			++FCC LINE CHARGE++	
			(DIRECT INCOMING TRUNK)	56.17
	1	TB2	/TER 212 555-1840: 002	
			/TA 12 MO, 06-20-90	
			/LSO 212 555	
			/CLT	
			96,DINA,212,555,1840,002	
			++FCC LINE CHARGE++	
			(DIRECT INCOMING TRUNK)	56.17
	1	TB2	/TER 212 555-1840: 003	
			/TA 12 MO, 06-20-90	
			/LSO 212 555	
			/CLT	
			96,DINA,212,555,1840,003	
			++FCC LINE CHARGE++	
			(DIRECT INCOMING TRUNK)	56.17
	1	TB2	/TER 212 555-1840: 004	
			/TA 12 MO, 06-20-90	
			/LSO 212 555	
			/CLT	
			96,DINA,212,555,1840,004	
			++FCC LINE CHARGE++	
			(DIRECT INCOMING TRUNK)	56.17
	1	TB2	/TER 212 555-1840: 005	
			/TA 12 MO, 06-20-90	
			/LSO 212 555	
			/CLT	
			96,DINA,212,555,1840,005	
			++FCC LINE CHARGE++	
			(DIRECT INCOMING TRUNK)	56.17
(7)	1	D1F2X	/TER 212 555-1840 001	
			(LOOP CHARGE 2-WIRE)	19.08
	1	D1F2X	/TER 212 555-1840 002	
			(LOOP CHARGE 2-WIRE)	19.08
	1	D1F2X	/TER 212 555-1840 003	
			(LOOP CHARGE 2-WIRE)	19.08
	1	D1F2X	/TER 212 555-1840 004	
			(LOOP CHARGE 2-WIRE)	19.08
	1	D1F2X	/TER 212 555-1840 005	
			(LOOP CHARGE 2-WIRE)	19.08
(8)	60	ND4	SLIDING SCALE CHARGE	10.92
(9)	5	9ZR	FCC LINE CHARGE	27.40

1. ZZYEB — USOC that tells you that this customer has 60 telephone numbers, 555-1840 through 1899. These numbers are reserved for use on their five DID trunks.

2. ND4 — DID numbers are grouped in groups of 20. Here the first group of 20 numbers are 555-1840 through 1859.

3. ND4 — DID numbers 555-1860 through 1879.

4. ND4 — DID numbers 555-1880 through 1899.

5. TB2 — Trunk Charge. DID trunks are identified by using the same telephone number as your main billing number, plus a suffix of 001, then 002 etc.

6. A DID is formatted in Circuit ID, telephone number format. Here the Circuit ID is 96,DINA,212,555,1840,001. The service code DI equates to "Direct-In Dial" as detailed on the service code chart (on pages 38&39). The modifier (position 5) "N" denotes that a DID trunk is used for a non-data (voice) purposes.

Each DID trunk is given a different suffix. The next DID trunk is identified as 96,DINA,212,555,1840,002 up to 96,DINA,212,555,1840,005.

7. Each DID has one loop charge associated with it.

8. The charge for 60 direct in station telephone numbers (555-1840 through 1899) is $10.92 or $3.94 for each group of 20 numbers.

9. Each DID also gets a FCC line charge. The FCC Line Charge is billed at a rate of $5.48 per line.

To determine the true monthly rate for any Bell company DID trunk, you must add up all the DID components. This includes loop charges and FCC Line Charges.

Most of the USOCs used by the various Bell companies are very similar. For example a DID from Pacific Bell is billed the following USOCs:

PACIFIC BELL DID USOCs

ND8 — Each group of 100 telephone numbers under first 200

NDA — Each group of 100 telephone numbers above the first 200

TMN — DID trunk charge- assured, measured rate

TCT — Circuit termination charge per DID trunk

9ZR — FCC line charge

The TMN USOC is equivalent to New York Telephone's TB2 USOC, while the ND8 USOC is the equivalent of New York

Telephone's NDZ USOC. The TCT USOC equates to New York Telephone's D1F2X USOC (2-Wire loop charge).

The following partial CSR from New Jersey Bell further illustrates the similarity in DID billing between the Bell companies:

USOC	DESCRIPTION	RATE
ND8	DID 1ST 20 LINE NUMBERS	$20.00
TDD	/TN 609 555-9999A	$8.00
NDT	/TN 609 555-9999A	$38.97
	/DID TRUNK CHARGE	
9ZR	/FCC LINE CHARGE	$3.82

On a New Jersey Bell CSR, a DID trunk is identified with a telephone number plus an alpha prefix (555-9999A) as opposed to the numeric prefix assigned on a New York Telephone CSR (555-9999 001).

T1 BILLING

A T1 is nothing more than a digital POTS line that is designed to handle 24 voice conversations simultaneously. Having a T1 is therefore equivalent to having the capacity of 24 POTS lines.

A T1 also uses two pairs of normal twisted wire [as described in our local loop (1MB & ALN) billing]. Voice (analog) conversations are digitized and then multiplexed so they can share the single line. Because a T1 is a digital line, it can also handle data traffic or any combination of 24 separate voice or data transmissions. One way to visual a T1 is to see it as a pipe that is subdivided into 24 separate channels.

A T1 can terminate at the CO, or it can be connected to another location, much like any other special service circuit. If, for example, you have 14 FDDA (full duplex-data) and 10 TLNA (tie line) circuits to a satellite location across town, these data lines can be consolidated on to a single T1. Chapter 3 shows how you can determine when, where and how much money can be saved by placing data lines on a T1. A T1 that connects two locations (other than a connection to the CO) is called a point to point T1.

A T1 is often used to connect voice lines directly to a long distance carriers POP (Point of Presence). This arrangement utilizes a point to point T1 that originates at your location and terminates at the POP. Chapter 3 illustrates how a T1 that connects you to a POP saves you money on long distance bills.

A T1 to your local CO is usually justified as a way to get a discount from the telephone company by "bulk purchasing" local loops. In this arrangement, a T1 will typically replace DID trunks. 24 DID trunks can be placed onto a T1 to your local CO. Chapter 3 details how this conversion can save you money.

To use a T1 to replace DID trunks you need to be served by a central office that utilizes a digital (as opposed to an analog) switch. Common digital switches utilized by the Bell companies include the DMS100, manufactured by Northern Telcom, and the No. 5 ESS manufactured by AT&T. Typical analog switches still in use include the AT&T No.1 ESS, No. 2 ESS and No. 3 ESS. To find out if your CO utilizes a digital switch, call the telephone company directly.

Pricing for a T1 favors densely populated areas like Manhattan, New York. It is costly to build the infrastructure to support digital facilities. Economies of scale favor cities and large office parks.

As previously detailed, mileage billing for standard special circuits usually contain two components: a fixed minimum charge and a variable mileage charge. Variable mileage or interoffice mileage is based on the distance between the COs serving each end of the circuit. A T1 is nothing more than an high capacity special service circuit.

Mileage charges billed on standard special service circuits, did not take into account the distance from each end of the circuit to its local CO. In densely populated, high use areas, there are many COs, and no one business tends to be too far from a CO. However, in suburban and rural areas, customers can be a significant distance from their CO. T1 mileage billing takes into account the distance from your business to the local CO. Many Bell companies determine mileage charges by calculating the distance to the local CO in 1/4 or 1/2 mile increments.

POINT TO POINT T1 BILLING

New York Telephone calls its point to point T1 service, SUPERPATH. The following summarizes the major monthly and installation charges of a SUPERPATH T1:

ITEM	MONTHLY CHARGES	INSTALLATION CHARGES
1. SERVICE ORDER		$56.00
2. TWO PREMISE VISIT CHARGES		$38.00
3. TWO CO LINE CHARGES		$142.70
4. LOCAL DISTRIBUTION CHANNEL FOR CKL-1		
A) FIXED-MINIMUM CHARGE	$269.93	$528.64
B) EACH ADDITIONAL 1/4 MILE TO CO OR FRACTION	$31.15	NO CHG.
5. LOCAL DISTRIBUTION CHARGE FOR CKL -2		
-A) FIXED MINIMUM CHARGE	$269.93	$528.64
B) EACH ADDITIONAL 1/4 MILE TO CO OR FRACTION	$31.15	NO CHG.
6. INTEROFFICE MILEAGE(INOF)		
A) FIXED MINIMUM CHARGE	$129.78	$1,585.94
B) EACH ADDITIONAL MILE	$41.53	NO CHG.

The total installation cost for a point to point SUPERPATH T1 is $2,879.92 regardless of the mileage involved. Two premise visits and two CO line charges apply because installation work has to be done at both ends of the T1 (CKL-1 and CKL-2). To calculate the monthly charge we have to know what the distance is between each location and its local CO. We also need to know the distance between the COs serving each location.

Lets analyze an condensed CSR for a SUPERPATH T1. Our first example details the billing for a T1 that directly connects two locations that transmit large volumes of data to each other.

Rates will often vary by areas within a Bell company. New

York Telephone has two sets of T1 rates for what they refer to as "downstate" (the New York City Metropolitan Area), and another set of rates in upstate New York. These rates are suppose to reflect the differences in the cost of doing business in these different areas. They also reflect the fact that New York Telephone faces greater competition in the New York City area. The following CSR details billing for a T1 within the New York City Metropolitan-Area (LATA 132):

FIGURE 2.13

A CSR FOR A SUPERPATH T1

	QUANT	ITEM	DESCRIPTION	RATE
(1)	1	HCAXS	/CLS 96,DHDA,999363,,NY (HIGH CAPACITY CIRCUIT)	
(2)	1	XUW1X	/CLS 96,DHDA,999363,,NY (MILEAGE CHARGE)	129.78
(3)	3	3LN1S	/CLS 96,DHDA,999363,,NY (MILEAGE CHARGE)	124.59
(4)		CKL	1-150 BROADWAY MANHATTAN, NY /CLS 96,DHDA,999363,,NY /LOC FLR 12/LSO 212 555 /TAR 002	
(5)		SN	GLOBAL NETWORKS INC	
(6)	1	XUN1X	(MILEAGE CHARGE)	269.93
(7)	1	XUD1Y	(MILEAGE CHARGE)	31.15
(8)		CKL	2-1299 AVENUE OF THE AMERICAS, MANHATTAN, NY/CLS 96,DHDA,999363,,NY /LOC FLR 6/LSO 212 999	
(9)		SN	BIG EIGHT INC	
(10)	1	XUN1X	(MILEAGE CHARGE)	269.93
(11)	1	XUD1Y	(MILEAGE CHARGE)	31.15

1. HCAXS (Circuit ID) — A T1 has a Circuit ID and can be identified by checking positions 3, 4, and 5 (as with a standard special service circuit). We see from our service code chart (on pages 38&39) that DK is the service code for digital data and the "D" in position 6 tells us it is used for data transmission. The Circuit ID is in serial number format. The following analizes a T1 Circuit ID:

PREFIX	SERVICE CODE	MODIFIER	SERIAL NUMBER
1 2	3 4	5 6	7 8 9 10 11 12
9 6	D H	D A	9 9 9 3 6 3

2. XUW1X — USOC that denotes the Interoffice Channel Charge. This charge represents the fixed mileage charge which is billed regardless of the distance between the COs serving each location.

3. 3LN1S — USOC that denotes the Variable Interoffice Mileage Charge. The quantity field tells us that the INOF mileage is 3 miles. The local and foreign COs are three miles apart. (As discussed earlier, when a USOCs' second character is L, as in 3LN1S, expect mileage billing). The monthly charge per 1/4 mile is $41.53 or $124.59 (3 x $41.53), for this customer.

4. CKL 1 — This USOC denotes the originating location, 150 Broadway, and also tells us that the customer is located on the 12th floor.

5. SN — USOC that denotes the name of the customer at the originating location. The customer at the originating location is Global Networks, Inc.

6. XUN1X — USOC that denotes the fixed Local Distribution Charge. The Local Distribution Channel (LDC) is the channel from your local CO to your premise. The fixed minimum cost for the channel is $269.93.

7. XUD1Y — USOC that denotes the Local Distribution Mileage Charge. This is where you are charged extra for the distance (in 1/4 miles) that you are physically located from the CO. The charge is $31.15 per 1/4 mile or fraction thereof. As this customer is located in densely populated Manhattan, the customer is within 1/4 mile of the CO, and is only billed an minimum mileage charge of $31.15.

8. CKL 2 — Our terminating location is at 1299 Avenue of the Americas, floor 6.

9. SN — The customer at the terminating location is Big Eight, Inc.

10 & 11. — Repeat of 6. & 7. for the terminating end.

Schematically, our customer is depicted as follows:

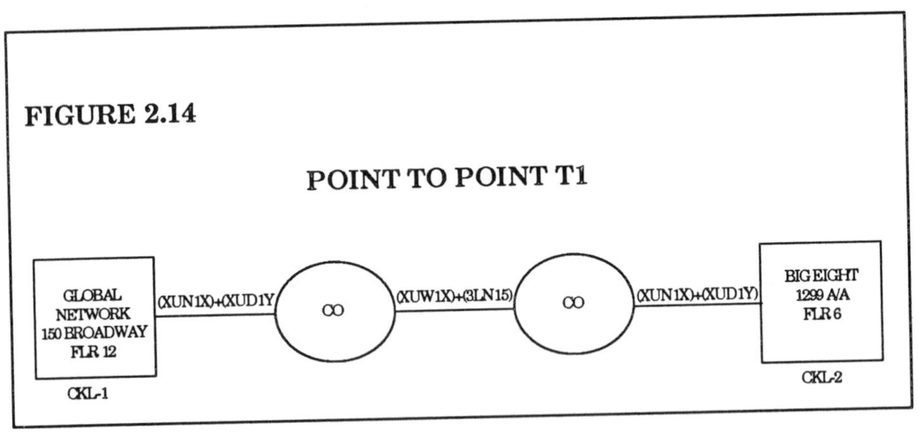

FIGURE 2.14

The monthly charges for this point to point T1 is determined by the following:

a) Both the originating and the terminating end of the T1 is within one 1/4 mile of their local COs.

b) The originating and the terminating COs are 3 miles apart.

We can summarize the monthly billing from the previous CSR as follows:

ITEM	USOC	MONTHLY
1) LOCAL DISTRIBUTION CHANNEL		
FOR CKL-1		
A) FIXED	XUN1X	$269.93
B) PER 1/4 MILE OR FRACTION	XUD1Y	$ 31.15
FOR CKL-2		
A) FIXED	XUN1X	$269.93
B) PER 1/4 MILE OR FRACTION	XUD1Y	$31.15
2) INTEROFFICE MILEAGE		
A) FIXED	XUW1X	$129.78
B) PER MILE OR FRACTION (THREE MILES TOTAL)	3LN1S	$124.59
TOTAL		$856.53

New Jersey Bell billing for T1s is simpler and cheaper than New York Telephone's procedure. New Jersey Bell will bill a point to point T1 on a CSR as follows (condensed CSR):

LIST SECTION

LST	ABC HOSPITAL
LA	101 NEW JERSEY TURNPIKE, NWK
CKL	(A) 12 BAYONNE ROAD
CKL B	(B) 234 12TH ST., PORT JERSEY

SERVICE AND EQUIPMENT SECTION

QUANT	ITEM	DESCRIPTION	RATE
1	CKT	DHSA 999041NJ	
1	DIF	1.544 MBPS	
1	HCAXS	/HI CAP CIRCUIT	
1	1ROP2	/CKL A/LSO 201 999	215.00
1	1ROP2	/CKL B/LSO 201 555	215.00
6	1LWPX	INTEROFFICE CHANNEL	292.00

The total cost of a point to point T1 in New Jersey that connects locations that are served by COs 6 miles apart is $722.

We can now see that the monthly cost of a point to point T1 that is 3 miles apart will cost $856.53 monthly from New York Telephone, while a point to point T1 that is 6 miles apart will cost $722 when provided by New Jersey Bell.

BILLING FOR A T1 TO THE LOCAL CO

Another type of T1 is a direct connection to your local CO. As discussed, this configuration is used to replace CO lines (POTS or DIDs). Chapter 3 will show you that the most cost efficient use of a T1 to the CO is to replace existing DID trunks.

New York Telephone calls its T1 service to the CO, FLEXPATH. You need a digital PBX, and must also be in a digital CO in order to take advantage of this offering. The pricing of this type of T1, is in part based on the assumption that at least some of your 24 ports will be used to replace DID trunks.

A FLEXPATH T1 circuit has the capacity to handle a total of 160 DID numbers. If a customer installs a FLEXPATH T1 with 160 DID numbers in Manhattan, he would pay the following monthly and installation charges: (Since only one location is involved, only one premise visit is required).

COST OF A NEW FLEXPATH T1 — MONTHLY AND INSTALLATION CHARGES:

ITEM	MONTHLY	INSTALLATION
SERVICE ORDER CHARGE		$56.00
PREMISE VISIT CHARGE		$19.00
FLEXPATH T1		
160 DID NUMBERS- 1ST GROUP OF 100 @ $18.23		
NEXT THREE GROUPS OF 20 @ $3.64 EACH	$29.15	
RENTAL OF 24 PORTS	$533.06	
DIGITAL TRANSPORT FACILITY		
a) FIRST 1/2 MILE OR FRACTION TO CO	$435.87	$1,550.00
b) EACH ADDITIONAL 1/2 MILE OR FRACTION TO CO	$269.86	______
TOTAL	$1,267.94	$1,625.00

We have previously identified the monthly cost of a POTS line in Manhattan as:

1MB	$16.23
TJB	$ 4.87
9ZR	$ 5.48
TOTAL	$ 26.58

At these rates it is not cost effective to put 24 voice lines at a monthly rate of $ 637.92 (24 x 26.58) on a FLEXPATH T1 that bills $998.08 monthly. Chapter 3 shows when, and for what services, a T1 becomes cost effective.

SAMPLE T1 BILLING

As with New York Telephone, Pacific Bell bills each point of the T1 one Local Distribution Channel (LDC). LDCs are classified as A through D and vary on price according to your configuration. Figure 2.15 is an example of how Pacific Bell bills a T1, served by COs 5 miles apart.

These examples should give you a feel for the general practices and procedures used by the local Bell companies to bill various T1 configurations. While there are many variations

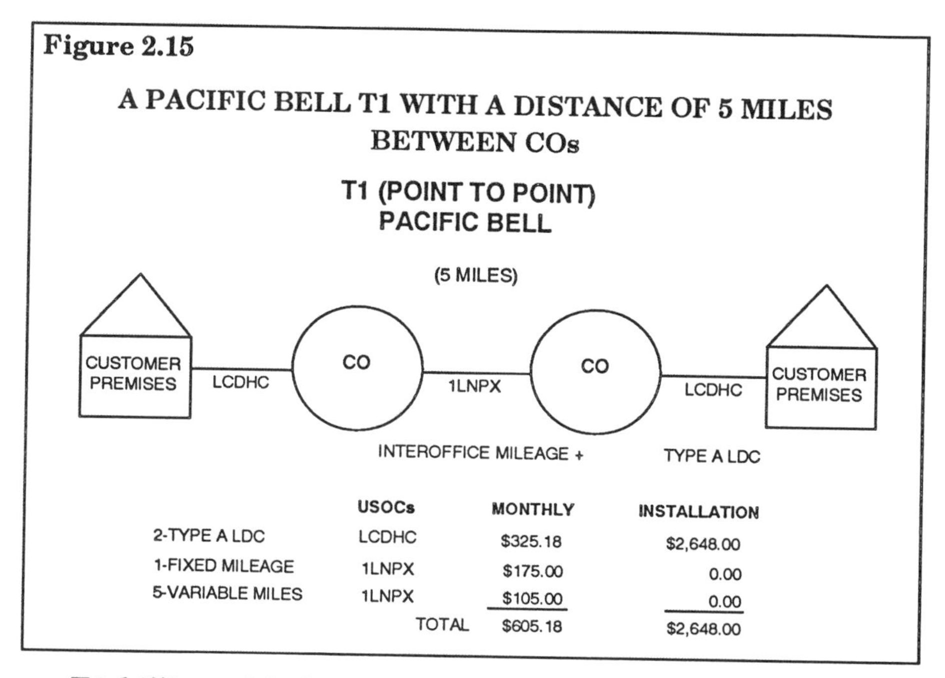

	USOCs	MONTHLY	INSTALLATION
2-TYPE A LDC	LCDHC	$325.18	$2,648.00
1-FIXED MILEAGE	1LNPX	$175.00	0.00
5-VARIABLE MILES	1LNPX	$105.00	0.00
	TOTAL	$605.18	$2,648.00

Figure 2.15

A PACIFIC BELL T1 WITH A DISTANCE OF 5 MILES BETWEEN COs

on T1 billing, this basic knowledge should help you identify T1s that appear on your CSR, and identify the locations that they connect.

SWITCHED 56 Kbps SERVICE

Rather than pay for a dedicated digital line between your locations, you may want to consider switched 56 Kbps service. It is fast, easy and has the potential to revolutionize the way data is sent. Switched 56 is a dial up line that is digital. Rather than pay for a dedicated digital line between locations, you pay only for the time you are connected, much like a voice call. Because you use the switched network, you can dial any location and send data at high speeds to that location. The catch? You need to install special equipment (CSU/DSU) at each location you wish to send data to, and also pay the local telephone company a monthly fee. The local company also charges you at a higher rate for these calls (as opposed to a similar voice call). Finally, all locations that you wish to call must be served by a digital CO. Figure 2.16 depicts a SWITCHWAY configuration.

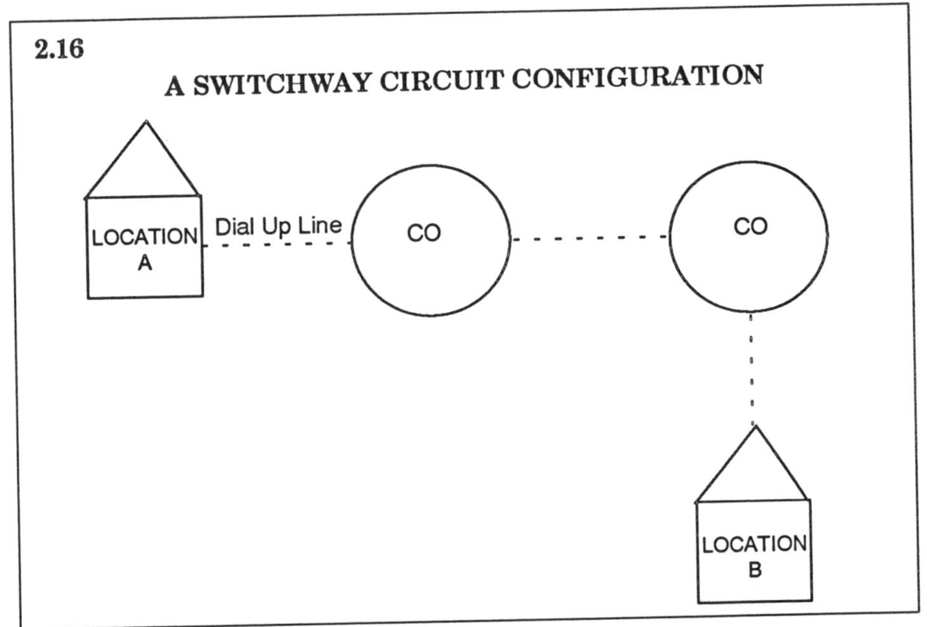

The following is an example of a customer using SWITCHWAY, New York Telephone's switched 56 service.

This customer has two locations in New York City. The charges per location are as follows:

Measured Data Line (2 or 4 Wire)	$32.18
Service Arrangement	$55.02
FCC Line Charge	$5.48
TOTAL	$92.68

Two locations: $185.36 total. Installation charges for each location will run $404.90 ($809.80 total).

Voice calls made within New York City are billed at a rate of $.08 for the first three minutes and $.013 each additional minute. Data calls made on a SWITCHWAY circuit are levied a surcharge as follows:

Initial Period (or fraction thereof):

1 minute	.07 cents
2 minutes	.15 cents
3 minutes	.22 cents
4 minutes	.29 cents
5 minutes	.36 cents

Each additional minute or fraction thereof over and above 5 minutes is billed at a rate of $.07. These charges are added to the normal voice rate charged by New York Telephone.

Example of how to price a 5 minute SWITCHWAY call between locations in New York City.

Rate the call based on existing voice call rates:

First three minutes	$.08
Next two minutes ($.013 per min.)	$.026
Subtotal	$.106
Then add SWITCHWAY surcharge:	$.360
Total Cost of Call	$.466

A 10 minute SWITCHWAY call within New York City will cost:

1ST 3 minutes	$.08
Next 7 minutes	$.091
SWITCHWAY surcharge	$.72
Total cost of call	$.891

Rate ccomparison between various Bell companies for switched 56 service (reprinted, in part, from Teleconnect Magazine):

RBOC	MONTHLY	INSTALLATION	PER MINUTE
AMERITECH SWITCHED DIGITAL SERVICE (SDS)	$107	$750	3 CENTS
BELL ATLANTIC SWITCHED 56 Kbps	$150	$725	14 CENTS
BELLSOUTH	$ 55	$ 55	12 CENTS 1ST MIN/10 CENTS ADD. MIN
PACIFIC BELL SDS	$ 45	$500	SAME AS VOICE RATES

SWITCHED 56 Kbps VERSUS DEDICATED 56 Kbps DIGITAL ACCESS

A cost comparison of switched 56 kbps and dedicated 56 kbps service largely depends on three factors:

1. Usage between locations

2. Distance between locations

3. Number of locations

The following compares the costs involved in utilizing SWITCHWAY (switched 56) as opposed to DIGIPATH service.

Dedicated Digital Service II (DDS II). Based on New York Telephone rates. Both locations are in New York City, 5 miles apart.

Dedicated Access (DDS II)

ITEM	MONTHLY	INSTALLATION
SERVICE ORDER		$56.00
2 PREMISE VISITS		$38.00
2 CHANNEL TERMINATIONS	$149.78	$1054.82
FIXED MILEAGE	$30.53	
5 MILES AT 4.91 EACH	$24.55	
TOTAL	$204.86	$1,148.82

Same customer installs SWITCHWAY at both locations:

ITEM	MONTHLY	INSTALLATION
SERVICE ORDER		$56.00
CO LINE INSTALLATION CHARGE ($71.75 PER LOCATION)		$143.50
2-PREMISE VISITS (19.00 PER LOCATION)		$38.00
2-SYSTEM ESTABLISHMENT CHARGE ($404.90 PER LOCATION)		$809.80
2- MEASURED DATA LINE CHARGE ($32.18 PER LOCATION) $	$64.36	
2- SERVICE ARRANGEMENT CHARGE $110.04 PER LOCATION)	$110.04	
2- FCC LINE CHARGES (5.48 PER LOCATION)	$11.86	
TOTAL	$186.26	$1,047.30

The installation charges for a dedicated line are $101.52 more than that for SWITCHWAY service. Monthly charges run $186.26 vs. $204.86 (the dedicated line is $18.60 more per month).

The main difference between the services is that with SWITCHWAY you pay additional charges for usage. To calculate the monthly break even point that makes a SWITCHWAY installation cost effective, you have to determine the minutes of usage required to make up the monthly difference of $18.60.

First, deduct the charge for a five minute SWITCHWAY call. A five minute SWITCHWAY call will cost you $.466. Subtract this number from $18.60. This will equal $18.134. Now, divide $18.134 by the additional cost per minute of a SWITCHWAY call ($.013 per minute plus $.07 surcharge). 18.134 divided by $.083 equals 218.48 minutes. Add this number to the 5 minute initial period and the break even point for monthly usage is: [219 minutes (218.48 rounded) plus 5 minute initial period] 224 minutes. If you use the line for more than 224 minutes (3.73 hours) a month, it is more cost effective to utilize a dedicated line between locations

SWITCHWAY is a product that is expensive if you have heavy usage between locations. It does, however, offer the potential to revolutionize data transmission. Imagine that all your locations and all the locations of your suppliers and customers have switched 56 capability. You would maximize your initial investment in switched 56. Instead of putting in high speed dedicated digital lines between all these locations you could send large amounts of data to whomever you wanted. It would be as easy as FAXing a document.

With more and more companies utilizing Electronic Data Interchange (EDI) to exchange information, switched 56 could well be the answer to the increasing need for companies to transmit larger and larger amounts of data. The major stumbling blocks are the per minute pricing policies of the local telephone companies, and the fact that your CO switch must be digital in order to take advantage of switched 56 service.

CONCLUSION

Many companies lose track of special service circuits that are installed throughout the years. Few people take the time to develop the expertise to read and decipher their CSR. Without this knowledge, telephone billing errors perpetuate themselves. Simply by being able to identify and diagram special service circuits you will identify common overbilling errors. Chapter 3 provides details on how to identify common billing errors and how to receive refunds for overpayments to the local telephone company.

CHAPTER 3

HOW TO OBTAIN REFUNDS AND REDUCE YOUR MONTHLY TELEPHONE BILL

Most errors are easy to spot once you take the time to read and decode your CSR. Checking the accuracy of your telephone bill does not have to be time intensive. The following guide to finding errors is both fast and accurate.

HOW TO IDENTIFY COMMON BILLING ERRORS

There is an easy way to find 90% of all mistakes on your telephone bills:

Step 1 — Write down all the telephone numbers that appear on your CSR.

Step 2 — Note all lines that have duplicate USOCs on them. For example, you may come across a telephone line that has 2 TTB (touchtone) charges billed to it. A telephone line can only have one touchtone charge per line. The following example illustrates a CSR with 2 TTB charges on line number /8022.

QUANT	ITEM	DESCRIPTION	ACTIVITY DATE	RATE
1	ALN	/TN 8022/PIC ATX/PCA SN, 12-18-91 ++FCC LINE CHARGE CHARGE++ (ADDITIONAL LINE)	12-20-91	16.23
1	TTB	/8022 (TOUCH-TONE) SERVICE	06-15-88	3.08
1	TTB	/8022 (TOUCH TONE) SERVICE	06-15-88	3.08

This customer is overpaying for one TTB. The "start date" of the incorrect billing can be determined by the Activity Date. We can safely say that overbilling for this charge has been in existence since at least 06-15-88.

The total number of touchtone charges billed to your account can never exceed the total number of telephone lines. Similarly, the total number of FCC line charges (USOC 9ZR) should also equal (and never exceed) the total number of telephone lines billed on the CSR.

Step 3 — Take special notice of any telephone line that has unique or different USOCs billed to it. For example, assume that your CSR lists 37 telephone lines. If 2 of the 37 lines have optional wire maintenance charges, a red flag should go up. As a general rule, check all deviations from the basic 1MB (or ALN), TTB and FCC Line Charge which comprise the charge for a standard telephone line.

Step 4 — Call each telephone number listed on the CSR.

The results will fall into five categories:

1) Those telephone numbers that are answered. Make sure you verify that the person answering the phone works for your company. We have come across numerous situations where telephone numbers are on a company's CSR but are actually being used by a different, unrelated company. In addition to over-paying for the monthly rental of this line, you are also paying for someone else's phone calls. In this case you would be due credit for both the fixed monthly cost of the line and for all calls made over the line.

If your company has moved in the last five years, check your CSR for RCA or RCF numbers. RCA and RCF are the USOCs for remote call forwarding telephone numbers. These lines are designed to automatically forward calls from the dialed number to another designated telephone number. This service is similar to call forwarding. The exception being that this service is programmed at the CO and stays in effect all the time. When companies move, they often keep some of their old telephone numbers active for a 3 to 6 month period. They use RCA numbers to forward calls to their new telephone number.

This service (and the associated billing for the service) is suppose to end at a pre-arranged date. Unless you check your CSR, billing for this service can continue for years. In reviewing the CSR of one of our clients located in New Jersey, we found the following:

ACT DATE	USOC	QUANT	DESCRIPTION	RATE
06-17-87	RCA	5	/TN 201 999-2990/LSO 908 555 /FSO 908 999-1000/ REMOTE CALL FWDING	72.60

The CSR lists a TN of 201-999-2990, yet our client was located in the 908 area code. When we dialed 201 999-2990 it was answered at 908 999-1000, our client's main number. Yet, the client did not recognize the 201 number listed on the CSR. Further investigation revealed that the customer had moved to their current location on 06-17-87 and that 201 999-2990 was their telephone number at the old location. The quantity (5) listed on the CSR meant that up to five simultaneous telephone calls could be forwarded from the old telephone number to the new one. Once we alerted the telephone company of this error, the customer was credited $72.60 (plus tax and interest) per month back to 12-17-87.

2) Those telephone numbers that are busy. Redial these numbers later. They will eventually fall into one of the other categories.

3) Telephone numbers that are not answered. Dial these telephone numbers again at different times of the day and on different days of the week.

Check unanswered telephone lines against the line numbers listed on your long distance bill. Has there been any usage on these lines over the last three months? Do you have any records indicating that these telephone lines should have been disconnected. Interview your co-workers to see if anyone can supply you with information on the lines in question. (Performing an audit is a lot like being a detective. You will get out of it what you put into it).

If you use a call accounting system, check your old reports for usage on the lines in question. You can also call your local telephone company for help. Unlike the long distance carriers, the local telephone companies do not supply you with a list of your local calls with your bill. You can, however, call the telephone company and have them check your Local Usage Detail list (LUDs).

Your LUDs itemize all your local calls, sorted by telephone line. Be careful how you word your request as there is an additional charge if you request a print of your LUD List (approximately 50 cents per page). As an alternative to obtaining the LUDs, you can simply request the telephone company representative to visually check your LUDs for usage on the telephone lines not recognized by you.

Before you decide to disconnect any telephone lines, check your alarm company to ensure that you do not disconnect alarm lines. It is also a good idea to circulate a notice that lists the telephone lines you plan to disconnect. Wait 2 to 3 weeks before you take action.

Disconnecting even a small number of unused lines can save a significant amount of money. For example, if you disconnect just five lines (in N.Y.), your company will save $123.95 ($24.79 x 5) plus tax each month or $1,487.40 each year. Over a 5 year period you will save $7,437.00.

4) Those numbers that give off a tone when dialed. These lines are most likely connected to a FAX or modem line.

5) Intercept recording that advises that the line has been disconnected. This situation occurs more than you may think. Many companies are paying for telephone lines that are not even in service. You certainly deserve a refund for billing of lines that have been disconnected.

Step 5 — Diagram the originating and terminating locations of each special service circuit billed to your account. Even if you are a small company, you may be surprised at the number of special service circuits on your CSR. Look for partial disconnects of special service circuits. As previously shown, every special service circuit must connect at least two locations. The following example (for a customer served by New York Telephone Company) illustrates how a partial disconnect of a circuit will appear on a CSR:

QUANT	ITEM	DESCRIPTION	ACT DATE	RATE
1	PFSAS	/CKT 96,BANA,,99916,,NY	09-22-87	
	CKL	1-444 BROADWAY, MANHATTAN, NY/DES FLR 3/LSO 212 966/CLS 96,BANA,,99916,,NY		
1	CON2X	/EAR (LOOP CHARGE 2-WIRE)	02-1-90	21.53
1	NRASQ	/DES 42A BLK HOLMES INSTR SOUTH STAIRWELL		
1	PMWV2	/EAR (BASIC 2-WIRE VOICE CIRCUIT)	02-01-90	7.07

An examination of this CSR reveals that only one end of the circuit is listed (CKL-1). All circuits must connect at least two locations. In this case, the circuit was installed for use by a specific burglar alarm company. When the telephone company disconnected the circuit they only stopped the billing of the terminating end (CKL 2). Billing for CKL-1 continued.

COMMON MILEAGE AND SPECIAL SERVICE CIRCUIT ERRORS

Common special service circuit errors include the billing of inter office tariff rates on circuits that are actually intra office circuits. An intra office special service circuit connects two points or locations that are served by the same CO. Inter office circuits connect points or locations that are served by different central offices. In addition to physical mileage charges, an inter office circuit is often charged higher rates for the local loop connection and transmission functions. In some cases, additional rate differences exist when circuits travel short distances and/or originate and terminate within adjacent city blocks.

To identify these errors, you need to understand the general rules for special service billing and then verify the specific charges against what is detailed in the tariffs. Once you get accustomed to the telephone company billing practice in a certain area, you will be able to verify charges simply by a quick review of the charges listed on the CSR.

If you are too busy to research specific tariffs, contact the telephone company and ask to speak with a telephone company special service billing specialist. Sometimes you may have to speak with more than one specialist to get a full explanation of charges. Even if you reach a knowledgeable representative, you need to know how to ask the right questions. Most of the representatives will not go out of their way to discover billing errors for you. (By relying on these specialists as your sole source of information, you also run the risk that you may bring their attention to an under billing problem).

To start your verification, you need to find out if the circuit in question originates and terminates within the same CO (intra office circuit). This can be accomplished by calling the telephone company and providing them with the first three digits of the POTS telephone number that is currently used at each location.

If the circuit is a intra office circuit, inter office mileage charges should not apply. Additionally, if the originating and terminating points of the circuit are in close proximity, you may not be responsible for loop charges (i.e. CON2X or CON4X type USOCs) to the CO. The following examples illustrates the differences in rates for a variety of special service circuit configurations. (New York Telephone rates utilized).

Example #1 — OSNA type circuit that connects a PBX to a phone at an outside location served by a different central office. Figure 3.1 depicts the configuration of this circuit.

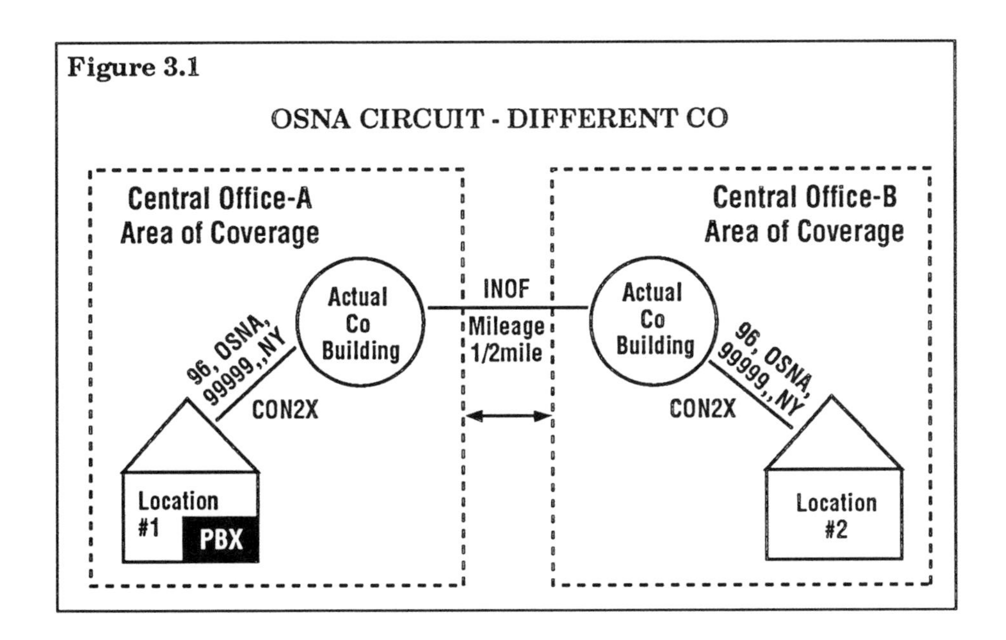

EXAMPLE #1 — Rates in Effect - Different COs

Qty	Item	Monthly	Installation
1	Service Order Charge		$ 56.00
2	Premise Visit Charges		$ 38.00
2	CO Line Connection Charges		$ 143.50
2	NRCCO (Mileage Termination Charge - For Non-Adjacent Blocks)		$ 318.00
1	Channel Connection Charge		$ 473.32
2	CON2X	$ 43.06	
2	Feature Function Charges - Type B (PMWFX)	$ 25.54	
	Mileage - Fixed	$ 36.86	
	Mileage - Variable ($3.17 per 1/4 mile)	$ 6.34	
	TOTAL	$111.80	$1,028.82

Example #2 — In this example the OSNA circuit both originates and terminates within the same central office area. Both locations are served by the same CO and are not located on adjacent city blocks. Since both locations are served by the same CO, monthly INOF mileage charges are not applicable. Similarly, installation charges for a channel connection does not apply. Diagram 3.2 illustrates how this circuit is designed.

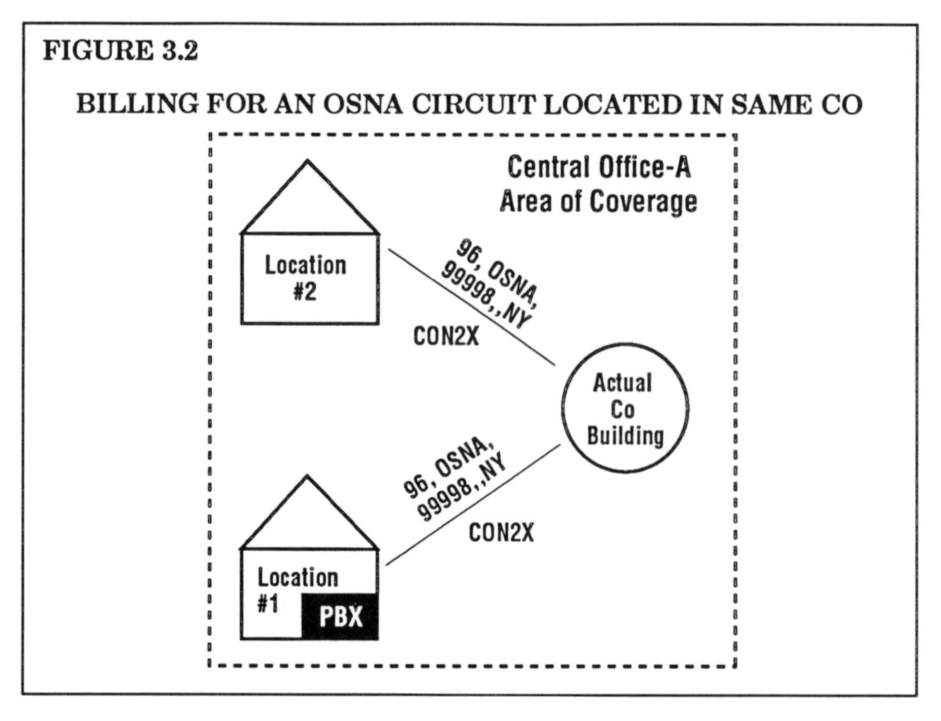

Example #2 — Rates in Effect - Same CO

Qty	Item	Monthly	Installation
1	Service Order Charge		$ 56.00
2	Premise Visit Charges		$ 38.00
2	CO Line Connections		$143.50
2	CO Line Connection Charges For Non-Adjacent Blocks)		$ 318.00
2	CON2X	$ 43.06	
2	Feature Function Charges — Type B (PMWFX)	$ 25.54	
	TOTAL	$ 68.60	$ 555.50

Example #3 — The monthly and installation rates to connect a OSNA circuit that is within the same CO, and is also located on an adjacent block, are even lower than the rates detailed in example #2. When you connect locations that are of close proximity, the circuit does not have to be connected via the local CO. Instead of paying 2 CON2X charges (one for each location connected to the CO) you pay one monthly BKP2X (block charge) to directly connect the locations. Your installation charges, therefore, do not include either the CO line charges or a channel connection charge since the circuit is not connected to the CO. Additionally, you only pay 2 NRCLF (mileage termination charge for adjacent blocks) at $131 per connection as compared to two NRCCO (mileage termination charge for non-adjacent blocks) charges at $159 per connection.

You should closely scrutinize all charges for circuits that connect locations that are in close proximity for overbilling. Figure 3.3 illustrates how this circuit is designed. The rates in effect for this type of circuit are itemized below.

Example #3 — Rates in Effect Adjacent Blocks

Qty	Item	Monthly	Installation
	Service Order Charge		$ 56.00
2	Premise Visit Charges		$ 38.00
2	NRCLF		$ 262.00
1	BKP2X	$ 27.73	
2 F	Feature Function Charges - Type B (PMWFX)	$ 25.54	
		$ 53.27	$ 356.00

FIGURE 3.3

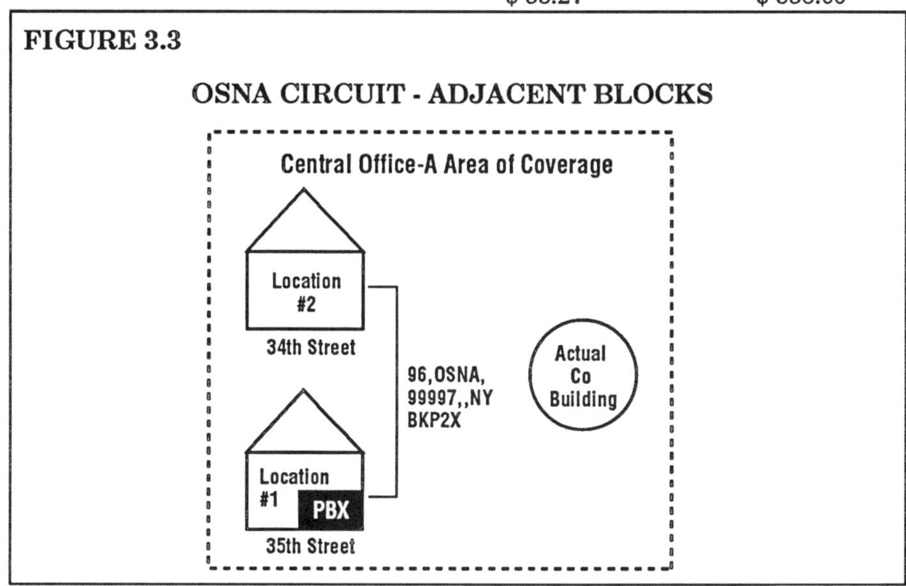

Example #4 — Details the monthly and installation charges on an inter office private line. Figure 3.4 illustrates how this circuit is designed.

Example #4 — Rates in Effect - Different COs

Qty	Item	Monthly	Installation
1	Service Order Charge		$ 56.00
2	Premise Visit Charges		$ 38.00
2	CO Line Charges		$ 143.50
2	NRCCO		$ 318.00
1	Channel Connection Charge		$ 473.32
2	CON2X	$ 43.06	
2	Feature Function Charges (Basic 2 Wire Voice)	$ 14.14	
	Autosignaling	$ 28.56	
	Mileage - Fixed	$ 36.86	
	Mileage - Variable ($3.17 per 1/4 mile)	$ 6.34	
	TOTAL	$ 128.96	$1,028.82

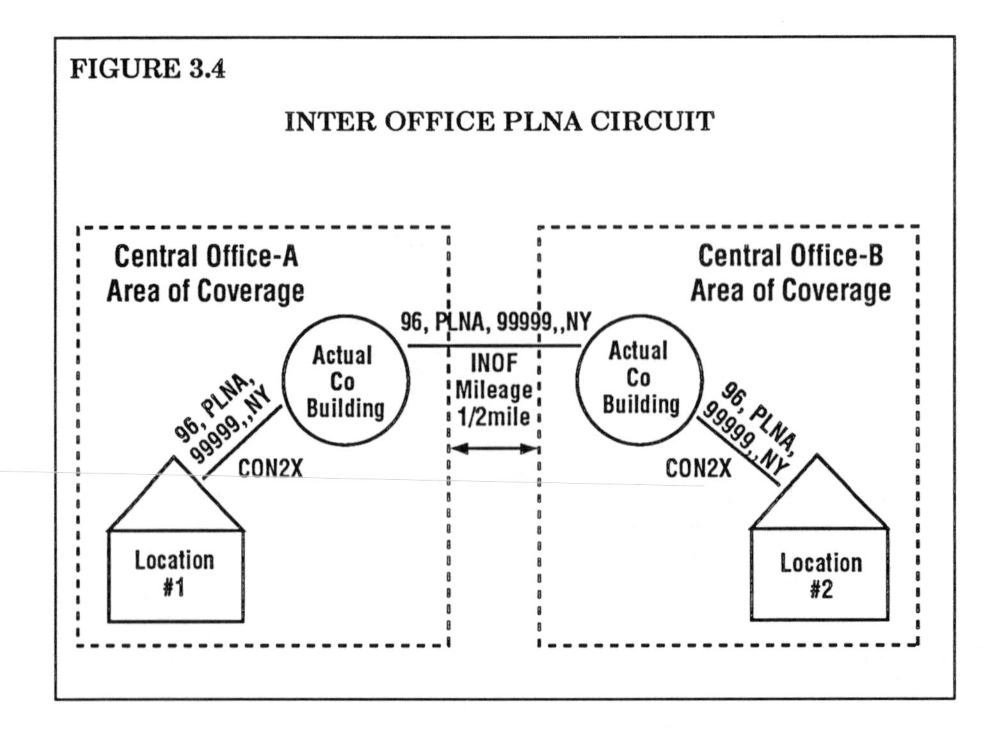

Example #5 — This example details the monthly and installation charges for an intra office private line. Figure 3.5 illustrates how this circuit is designed.

Example #5 — Rates in Effect - Same CO.

Qty	Item	Monthly	Installation
1	Service Order Charge		$ 56.00
2	Premise Visit Charges		$ 38.00
2	CO Line Charges		$ 143.50
2	NRCCO		$ 318.00
2	CON2X	$ 43.06	
2	Feature Function Charges (Basic 2 Wire Voice)	$ 1 4.14	
	Autosignaling	$ 28.56	
	TOTAL	$ 85.76	$ 555.50

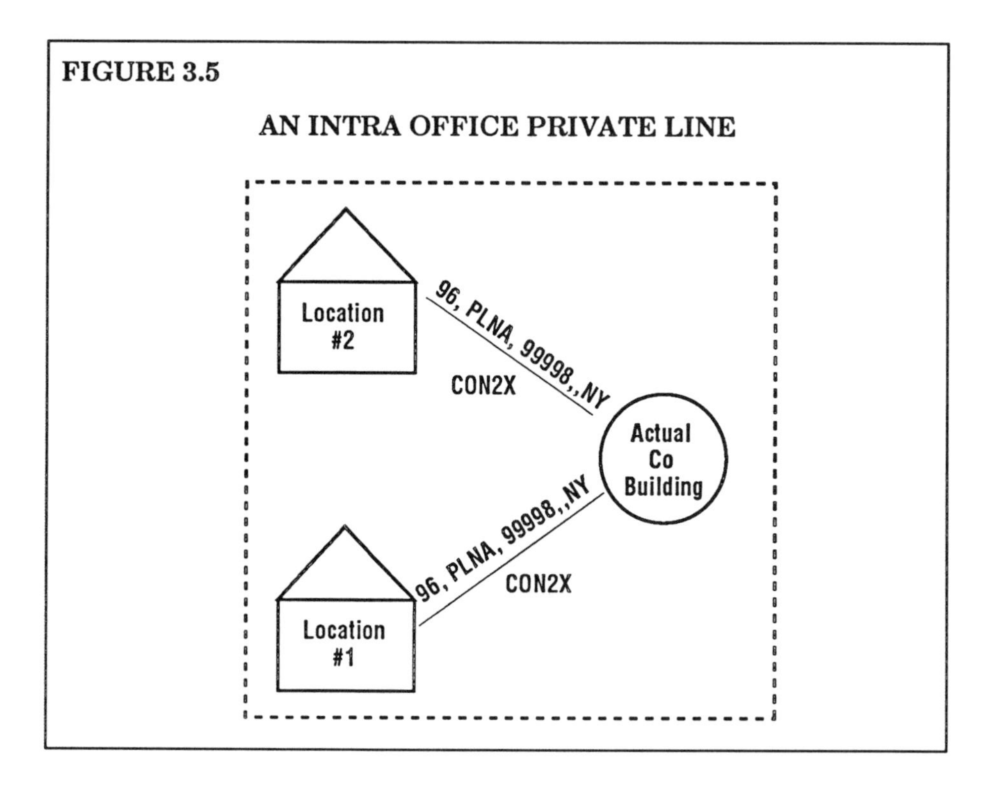

FIGURE 3.5

AN INTRA OFFICE PRIVATE LINE

There are also differences in rates between tie lines that are used to connect locations that have Centrex service as opposed to tie lines used to connect PBXs.

Examples 6 & 7 detail the differences in rates when Centrex service is involved.

Example # 6 — Monthly and installation charges for an inter office PBX to PBX tie line (4-Wire connection).

Example #6 — Rates in Effect Different COs.

Qty.	Item	Monthly	Installation
	Service Order Charge		$ 56.00
2	Premise Visit Charges		$ 38.00
2	CO Line Charges		$ 143.50
2	NRCCO		$ 318.00
	Channel Connection Charge		$ 473.32
2	CON4X	$ 81.22	
2	Feature Function Charges (Basic 4 Wire Voice)	$ 39.98	
	Mileage - Fixed	$ 36.86	
	Mileage - Variable-1/2 mile ($3.17 per 1/4 mile)	$ 6.34	
	TOTAL	$ 164.40	$1,028.82

Example #7 — Monthly and installation charges for an inter office Centrex tie line (2-Wire connection) as billed by New York Telephone.

Example #7 — Rates in Effect - Different C0s.

Qty	Item	Monthly	Installation
1	Service Order Charge		$ 56.00
2	CO Line Charges		$ 143.50
1	Channel Connection Charge		$ 473.32
2	Terminal Charges	$384.88	$ 527.86
2	Feature Function Charges (Basic 2 Wire Voice)	$ 14.14	
	Mileage - fixed	$ 36.86	
	Mileage - Variable - 1/2 mile ($3.17 per 1/4 mile)	$ 6.34	
TOTAL		$ 442.22	$1,200.68

One of the major billing differences between a Centrex tie line, and that ordered to connect PBXs, is the absence of premise visit charges. With Centrex service, your CO is, in essence, your PBX. Therefore, all the installation and connection activity takes place at the CO as opposed to your premises. You also do not have to pay for NRCCO (Mileage Service Termination Charge - for non-adjacent blocks) charges, but you do have to pay additional installation and monthly charges for CO terminal equipment. Although you do not pay for CON2X loop charges, Centrex monthly line charges apply. These charges are based on a variety of factors (dependant on whether your CO utilizes an analog or digital switch), and is not included in these rates.

Some of the local Bell telephone companies add additional complexity to their billing of an inter vs. an intra office circuit. New Jersey Bell, for example, varies the monthly rate for the local loop to the CO dependant on whether the circuit is designed as an intra or inter office circuit. The following examples detail differences in intra and inter office billing, as they appear on New Jersey Bell CSRs.

Example #8 — Intra office PXOS (PBX outside extension) circuit. Monthly charges as billed by New Jersey Bell. This example details the billing as it appears on a New Jersey Bell CSR. Figure 3.6 details how this circuit is designed.

FIGURE 3.6

NEW JERSEY BELL INTRA OFFICE OSNA CIRCUIT

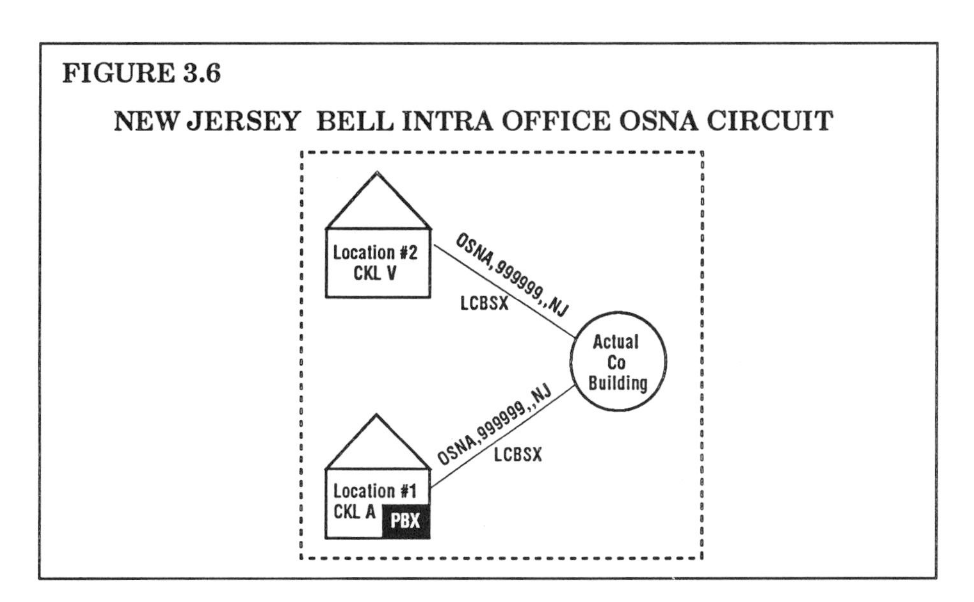

USOC	DESCRIPTION	MONTHLY RATE
CKT	OSNA 999999NJ	
ARR	PX 3XAYS/CKL A, V	
	SAY /** TYPE C SIGNALING ARRANGEMENT	.15
	LCBSX /CKL A/LSO 609 999/** LOC CHAN SM	

	EXCH	21.73
	LCBSX /CKL V/ LSO 609 555/** LOC CHAN SM EXCH	21.73
	TOTAL	$43.61

Example #9 — If the OSNA type circuit connects locations served by different COs (inter office) New Jersey Bell will bill as follows (COs are six miles apart in this example):

USOC/QTY	DESCRIPTION	MONTHLY RATE
CKT	OSNA 999998NJ	
ARR	PX 4XAYS/CKL A, M SAY /** TYPE C SIGNALING ARRANGEMENT	.15
	9B1SX / CKL A/LSO 609 999 / ** LOC CHAN DIF EXCH	23.46
	9B1SX / CKL M/LSO 609 999 / ** LOC CHAN DIF EXCH	23.46
6	1LVA4 /SEC 1/ EX WAYNE MILEAGE INTEREXCHANGE	28.42
	TOTAL	75.49

The monthly cost for this circuit totals $75.49. Figure 3.7 illustrates how this circuit is designed.

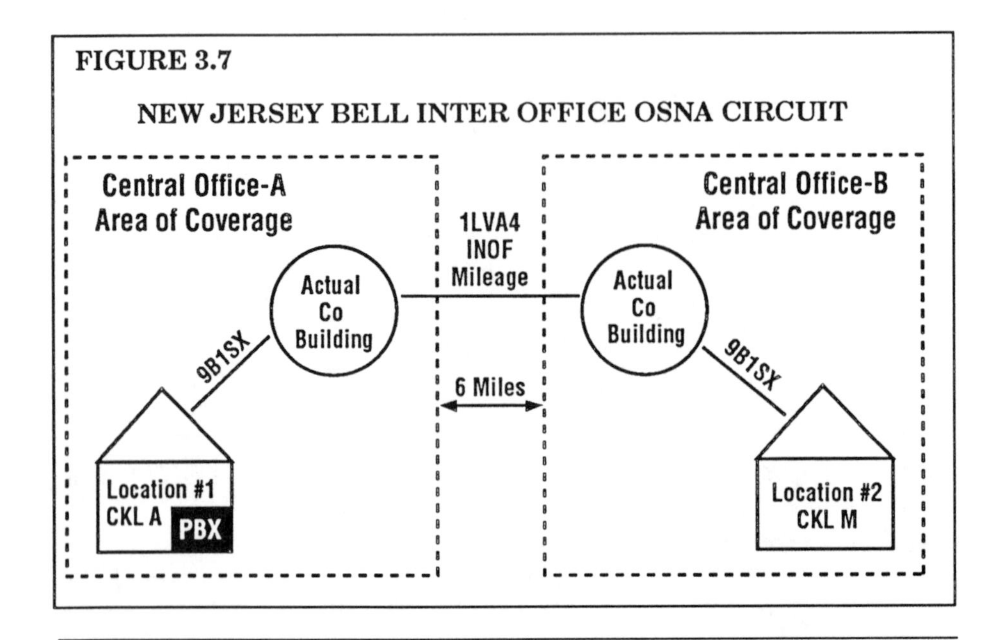

Example # 10 — If you connect 2 PBXs (via a tie line) located in the same CO, New Jersey Bell will bill you as follows:

USOC	DESCRIPTION	MONTHLY RATE
CKT	TLNA 999999NJ	
ARR	PX 3XAYS/CKL A, B	
	SLM /CKL A /** E&M TYPE	.90
	1SESP/ CKL A/LSO 609 999 / ** LOC CHAN SM EXCH	27.64
	SLM /CKL A /** E&M TYPE	.90
	1SESP/ CKL A/LSO 609 999 / ** LOC CHAN SM EXCH	27.64
	TOTAL	57.08

The monthly cost of this circuit totals $57.08. Figure 3.8 illustrates how this circuit is designed.

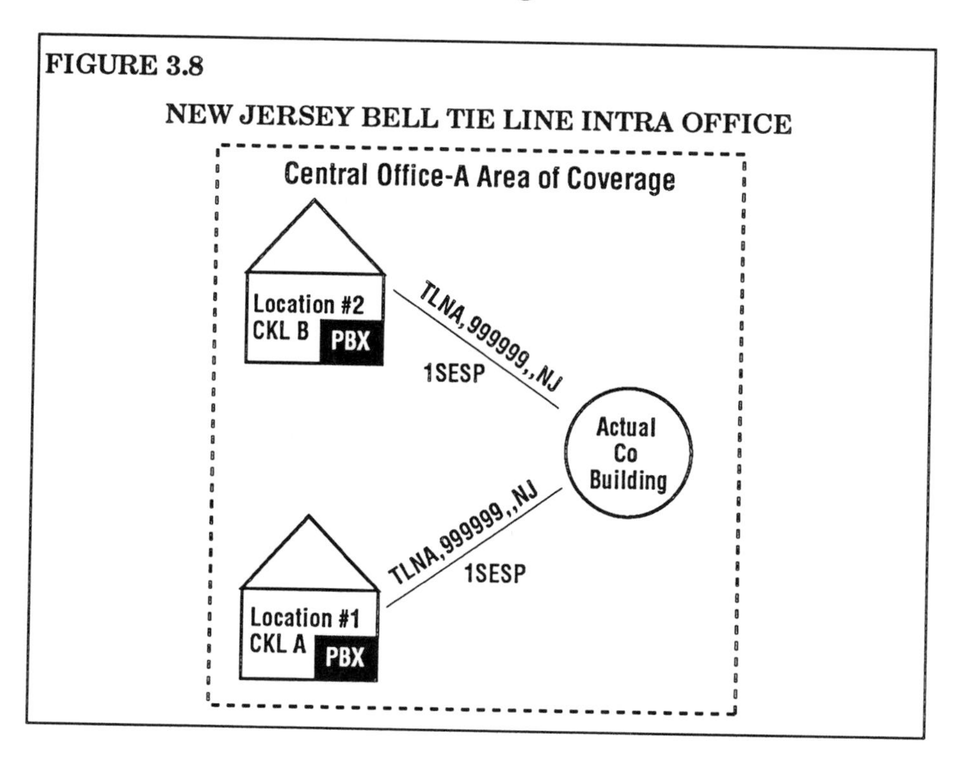

Example #11 — If you connect 2 PBXs located in different COs, New Jersey Bell will bill you as follows:

USOC/	DESCRIPTION	MONTHLY RATE
CKT	TLNC 999998NJ	
ARR	PX 3XAYS/CKL A, C	
	SLM /CKL A /** E&M TYPE	.90
	9B1SP/ CKL A/LSO 609 999 / **	
	LOC CHAN DIF EXCH	29.08
	SLM /CKL A /** E&M TYPE	.90
	9B1SP/ CKL C/LSO 609 999 / **	
	LOC CHAN DIF EXCH	29.08
6	1LTA4 / SEC 1/ EX WAYNE	
	INTEROFFICE MILEAGE	28.42
	TOTAL	88.38

The monthly cost of this circuit totals $88.38. Figure 3.9 depicts how this circuit is designed.

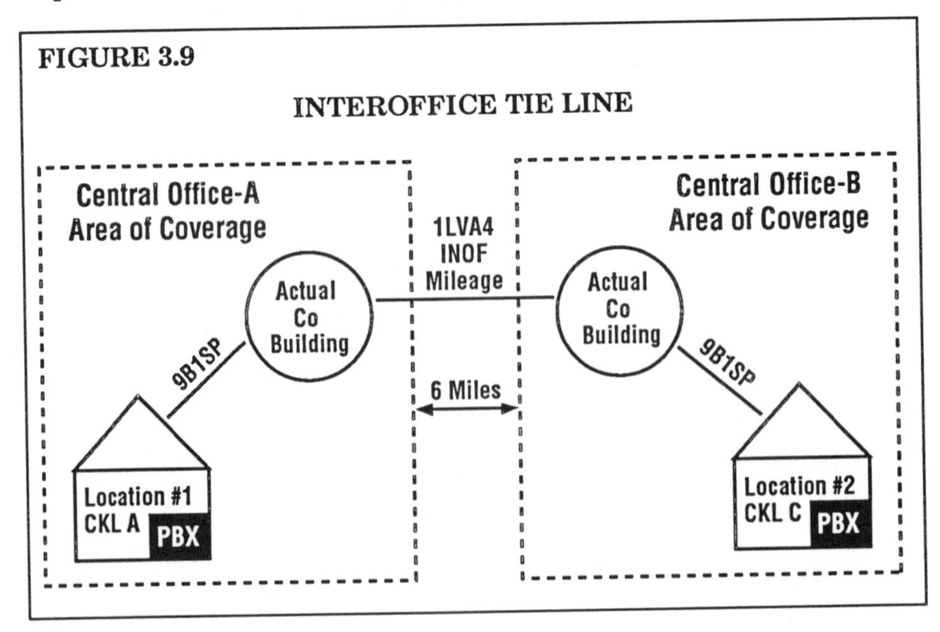

FIGURE 3.9

INTEROFFICE TIE LINE

There are many more examples of the different ways rates vary according to a circuit's design and function. To try and remember or document all of these differences would take quite a bit of time. You should, however, be aware that differences exist. Your knowledge should be broad enough so you can spot billing combinations that do not look right. Once you identify abnormalities, you can then look up the rates in question to determine its accuracy.

You should also be aware that some local telephone companies charge different inter office mileage rates based on the speed of the circuit provided. For example, most of the circuits supplied by New Jersey Bell are billed at a rate of $11.67 for the first mile and $3.35 for each additional mile. They are identified by the USOC code 1LTA4. However, circuits that are used as burglar alarms (low speed circuits) will be billed at a rate of $7.84 for the first mile and $6.18 for each additional mile. These mileage rates are identified by the USOC code 1L3A4.

As part of your verification of special service charges, contact your equipment vendor to verify requirements for feature functions (Type A, Type B or Type C PXOS charges) on each of your circuits. The billing of the wrong feature function charge is a common error.

TAXES

A billing error that affects more companies than you may realize is the incorrect billing of taxes. This error includes both the billing of state and local sales taxes, and the billing of federal excise taxes (a tariff of 3% on telecommunications services).

For example, common carriers, telephone and telegraph companies, radio and television broadcasting stations and networks are exempt from federal excise tax under Section 4253 (f) of the Internal Revenue Tax Code on their WATs service. Many trucking and shipping companies qualify as common carriers and are paying this tax in error.

Other exemptions to the federal excise tax include:

A. Charges paid by a Governmental Tax Exempt Organization.

B. Nonprofit educational organizations.

C. Schools operated as an activity of a church, parish or

other religious body.

D. Non profit hospitals.

E. Consular offices of a foreign government.

F. Community organization conducting the community action portion of an economic opportunity program.

Dedicated private network services are also exempt from this tax.

Exemptions for state and local taxes vary by state, but many of the federal exemptions apply.

Incorrect billing for taxes is not as obvious to spot as it may first seem. The billing of taxes is driven by the USOC codes that appear in the BILL Section of your CSR. For example, a New York Telephone customer that is federal, state and local tax exempt will have a BILL section that appears as follows:

ITEM	**DESCRIPTION**
BN1	RC HOSPITAL
BN2	HEALTH
BA1	781 E 42 ST
PO	NY NY 10036
TAR	002
TAX	FEDTELOCTE

The notation "FEDTELOCTE" after the USOC TAX tells us this customer is both federal and state and local tax exempt. The TE at the end of FEDTELOCTE denotes this customer is tax exempt for federal and local taxes.

This tax exempt indicator will also drive the billing of taxes on long distance calls that appear on your local bill as a service provided to the long distance carriers. You should also check your long distance bills that are separately provided by the long distance carriers for incorrect taxes.

Polices on refunds of taxes vary by telephone company. New York Telephone, for example, will refund up to three years of federal taxes, but only 3 months of state and local sales tax paid in error. New Jersey Bell will not refund any back taxes to you, while AT&T will refund taxes paid over the last two years. These companies will require you to send them proof of your exemption before they will refund taxes to you, and stop the future billing of taxes.

In any case, you can always get back taxes paid in error directly from the federal government by filing IRS form 843.

The IRS will refund taxes paid in error over the last three years. They require a log that details the telephone taxes paid, and you must keep your old telephone bills accessible in case they audit your refund request.

To recover back state and local taxes you can file with your local state tax agency. In New Jersey, for example, you can file form A-3730 to obtain a tax refund. State forms and refund policies vary state by state, but three years backcredit is common.

In all cases where you discover the incorrect billing of taxes, make sure you notify the telephone company to prevent future billing of taxes. Follow-up and check all future bills.

SURCHARGES

Surcharges are not considered taxes, and even tax exempt organizations must pay them. Adding surcharges to your telephone bill is one way state legislators raise revenue without technically raising taxes. Many surcharges are originally passed by state legislators as temporary revenue measures. More often than not they stay on your bill - indefinitely.

INITIATING AN OVERBILLING CLAIM

Once you identify an overbilling error, what is the next step?

1. Gather all the documentation you have. Check with your equipment vendor to obtain information they may have on moves and disconnects.

2. Try to determine the date the circuit or telephone line in question was disconnected. Be careful here. The activity date listed on the CSR may not be the date the circuit should have been disconnected. The activity date field will change

anytime a rate change is done that effects its associated USOC. Many times, telephone company employees will tell you that a certain circuit could not have been disconnected on a certain date. As proof they will point to the activity date and say someone must have requested a change to this circuit.

Try to find associated USOCs that are not billed a monthly rate to establish your case. The following example illustrates how this can be accomplished:

QUANT	ITEM	DESCRIPTION	ACT DATE	RATE
1	PFSAS	/CKT 96,BANA,,99916,,NY	09-22-87	
	CKL	1-444 BROADWAY, MANHATTAN, NY/DES FLR 3/LSO 212 966/CLS 96,BANA,,99916,,NY		
1	CON2X	/EAR (LOOP CHARGE 2-WIRE)	02-1-90	21.53
1	NRASQ	/DES 42A BLK HOLMES INSTR SOUTH STAIRWELL		
1	PMWV2	/EAR (BASIC 2-WIRE VOICE CIRCUIT)	02-01-90	7.07

In this case, the customer told us that this circuit was disconnected on 9-22-87. When we initially contacted the telephone company, their representative responded that this was impossible. The reason given was that the ACT DATE (activity date) listed on the CSR was 2-1-90. What the representative did not know is that New York Telephone had a massive rate change on 02-1-90 that effected most special service circuits.

We established that the USOC PFSAS (which lists the Circuit ID) was a better indicator of the last time physical activity occurred on this circuit. As there is not a monthly rate for this USOC, it is not effected by rate changes. The moral? Do not assume information given by telephone company representatives is always correct. Some representatives are better trained and more knowledgeable than others.

The following example shows how a typical overbilling case will progress:

Assume that after diagramming the locations of all your circuits you find that you are still being billed for off premise outside extensions (OSNA circuits) to a branch location that closed down five years ago. Your problem is that you can't find any records that prove you called the telephone company to disconnect them. What do you do?

Step 1 — Gather together all the information you have on the move. Look for anything that backs up your position and establishes the date you moved out of your branch office. For example, you may be able to get a copy of your old lease from your real estate agent or landlord. You may be able to locate moving bills from a trucking company which detail the date your vacated the premise in question. Look for any bill, record or lease that logically proves you were unable to use the circuit because you did not have access to the building the circuit terminates at.

Step 2 — Call the telephone company billing office and identify the circuits in question. Tell them the date you moved out of the building in question. Be prepared to overcome objections such as "how were we supposed to know that you moved?" Be persistent! Request a supervisor if you feel that you are getting the third degree. Your case will probably be sent to a specialist in overbilling claims. These special representatives investigate and correct overbilling problems. Put your claim in writing and send it to the specialist that is assigned to your case.

Step 3 — The telephone company will dispatch an installer or repair person to verify that the circuit is no longer in service (known by the telephone company as a physical inspection). Here is the sticky part. They will test the circuit to see if it is still able to function by putting a tone generator on the originating end and will then check the terminating end to see if they can pick up the tone. The logic behind this test is to determine if the circuit is usable.

If the circuit is still usable they will require proof that you requested a disconnect of the circuit. If the circuit is unusable, you will receive backcredit, but must work with the telephone company to determine the date the circuit should have been disconnected on. At this point, you will be required to negotiate with the telephone company. The stronger the case you present, the better your chances for success.

Step 4 — In about a week or two, you will get a call from the telephone company with the official results of their physical inspection. If the circuit is not working they will probably offer you 2 years back credit. If you are unhappy with what is offered, you do not have to take this initial offer as gospel. There is usually room for negotiation with the telephone company. If the circuit is working they will usually deny back credit.

THE SECRETS OF NEGOTIATING REFUNDS

Legally, the telephone company has a strong position when they deny you back credit. Many states have statue of limitation laws that limit the number of years of credit a company is required to give, even when its bills are proven wrong.

State regulatory commissions usually rule that it is the customer's obligation to review their telephone bill before paying them. The fact that you pay the bill is used as evidence that it is accepted by you as correct.

That said, you should realize that just because a company does not have to give you a refund to correct an error does not mean they can not. Many companies routinely issue credits to satisfy customers, and thereby increase good will. The logic used is that the customer is always right.

Think of how you would treat a customer who called in to complain about an overcharge on a bill sent by your company. If you are like most businesses and value your customers, you'll apologize and gladly correct the charge.

The telephone company, while still a monopoly, realizes it is in their best long term interest to verify and correct their bills. The telephone company, as a regulated entity, is also aware that good will is crucial to their bottom line. Rate hikes must approved by elected officials. Complaints about their bills does not help them gain approval for rate hikes.

Our advice is to be persistent. You need to speak to an executive that understands the bottom line and has the authority to help you.

The overbilling representative that denies credit is at the bottom of the telephone company totem pole. The hierarchy of a typical Bell telephone company is as follows:

Level	Job Title
Non-Management	Representative
First	Supervisor
Second	Manager
Third	District Manager
Fourth	Division Manager
Fifth	Asst. Vice President
Sixth (officer)	Vice President
Seventh (officer)	Chief Operating Officer
Eighth (officer)	President

The representative operates under strict guidelines. They can only offer credit in certain situations. They are also limited as to the amount of a refund that they can offer.

If you are not satisfied with the credit offered by the representative, ask to speak with a supervisor. Each level at the phone company has more leeway than the previous level to grant credit. You have more of a chance to present the logic of your case to the higher levels of management.

Document the strengths of your case in writing. Clearly state all your evidence, and why the denial of your request is unfair. In the case of the partial disconnect of the burglar alarm line, we had the burglar alarm company confirm in writing the date they no longer serviced my client.

Keep in mind that the telephone company is extremely level conscious. Employees will become increasingly uneasy as you ask for the next higher level manager. Officers at the telephone company are treated like living gods. Many times this strategy will prod the telephone company into making a business decision that grants your request for backcredit.

CONTACTING THE STATE PUBLIC UTILITY COMMISSIONS

You should only refer problems and disputes to State Regulatory Agencies as a last resort. Before you lodge a complaint you should at least take your case to the sixth level manager (officer level) at the telephone company.

Make it clear that you are planning to lodge a formal complaint. It will help your case.

Most local telephone companies have set up executive hotline numbers for their customers. For example, New York Telephone has set up what it calls the President's Help Line (800 722-2300). A call to this number usually generates results.

Many Public Regulatory bodies are moving toward systems whereby approval of rate hikes are dependant on the quality of service provided to telephone company customers. Because of this trend many telephone executives are rated on how many public complaints are registered against their department. These ratings generate their year end bonuses.

If your claim is denied by an upper level telephone manager, consider writing a letter to your local state senator. They can review your case and act as an intermediary. This allows the telephone company "an out". They can justify the credit in the context of the goodwill it creates.

If all else fails, contact your State Public Utility Agency. Once you make a complaint to them, their decision is final. Unfortunately, you will probably be disappointed at the results of your complaint. Their rules and regulations are usually quite explicit and Public Utility Agencies traditionally rule that it is the consumers' responsibility to question their bill before paying it. They will usually rule that simply by paying your monthly bill, you are accepting it as correct.

If you are still unsatisfied, consult a lawyer. This option is quite expensive and time intensive, and should only be considered when extremely large amounts of money are involved.

UNDERSTANDING YOUR REFUND

Once you agree with the telephone company on the date that the overbilling began, you now must decide if you want a refund check or wish to leave the credit on the bill.

Credits for overbilling are calculated by an automated billing system that tracks the dates different rates were in effect.

For example, a 1MB was not always $16.23. If the overbilling was in effect for five years your credit will take into account different rates and tax schedules in effect throughout the over-billing period. The following example illustrates this process:

Case 1: The telephone company technician confirms that a special service voice circuit on your bill is not operational. You provide the telephone company with a lease that clearly states that your company vacated the premises this circuit connects, back on June 25, 1986. The telephone company accepts June 25, 1986 as the date overbilling began. A credit is issued and you are advised that your credit will appear on the next bill (July 16, 1992). (You have the option to have this credit refunded to you via a separate refund check).

The following is an condensed version of a New York Telephone Company bill that itemizes overbilling credit.

Account Number: 212 999-5400 179
July 16, 1992
Page 1

Summary of New York Telephone charges

	Basic service July 16 through August 15	$157.34	
(1)	Service order/other charges and credits	9,857.92	CR
	Directory advertising	30.50	
	Local calls	57.01	
	Directory information	2.93	
	County emergency services surcharge	2.10	
(2)	Federal Tax (3%)	13.28	CR
(3)	State and Local taxes (8.25%)	665.06	CR
	Total	$10,286.38	CR

Basic service

(4)	These charges are for July 16 through August 15	$120.99
	Line Charge ordered by the FCC	30.90
	Wire maintenance charge (Optional)	5.45
	Total	$157.34

Page 2

Service order charges and credits

		Per month	Amount
(5)	Order R1AB9999 Feb 01, 1992 through Jul 15, 1992 CKT 96,CSV,999240		
(6)	CR 2 BASIC 2-WIRE VOICE CIRCUIT	14.14	$77.30 CR
(7)	CR 2 LOOP CHARGE 2-WIRE	43.06	235.40 CR
(8)	CR 011 MILEAGE CHARGE	71.73	392.12 CR
	Sub total		$704.82 CR

Other charges and credits

PON R9AB6669		
(9) Jun 25, 1986 through Jan 31,1992		
(10) CR 2 LOOP CHARGE		$487.73 CR
(11)CR BASIC 2-WIRE CIRCUIT		334.56 CR
(12) CR 11 MILEAGE CHARGE		6,096.88 CR
(13) INTEREST ON ADJUSTMENT		1,579.35 CR
	Sub total	$8,498.52 CR

Jul 01, 1992 through Jul 15, 1992		
(14) RATE ADJUSTMENT		$.13
(15) RATE CHANGE FCC LINE CHARGE		.99 CR
(16) NY GROSS INCOME TAX SURCHARGE		.05 CR
(17) MUNICIPAL SURCHARGE		170.32 CR
(18) N.Y. STATE/MTA SURCHARGE		483.35 CR
	Subtotal	$654.58 CR

1. — Your total backcredit for overbilling appears on the service order/other charges and credits portion of your telephone bill. The total listed credit is $9,857.92.

2. — Credit is given for federal taxes paid throughout the overbilling. The amount of credit is low because special service lines are generally exempt from federal excise tax. The credit here reflects federal tax charged on surcharges applied to the special service circuit.

3. — Credit for state and local taxes paid over the years is applied here.

4. — Your basic monthly service charge should be reduced from your previous bill in the amount of the monthly billing for the special service circuit in question (see items 6, 7 and 8).

5. — You will note that an order number (R1AB9999) appears near item 5. Order numbers that begin with "R" stand for record orders. This denotes that physical work is not required to disconnect this circuit. The order bypasses the provisioning and plant departments and is used to correct records only.

6, 7 & 8. — As previously discussed, uncovering billing errors will often lead to refunds, but they should always lead to a reduction in future bills. Your monthly reduction and corresponding credit for the period February 1, 1992 through July 15th is detailed here. It is extremely important to verify that your bill is reduced by the proper amount. Specifically, this bill is reduced by the following:

A)	2-Basic 2-Wire Voice Circuit Charges (7.07 each)	14.14
B)	2-Loop Charges (21.53 each)	43.06
C)	11-(1/4 miles) inter office mileage	71.73

You can verify that the correct monthly special service reductions are given by verifying these credits against the CSR used to uncover the billing error.

9. through 13 — Credit for the period June 25, 1986 through January 31, 1992 is detailed here for all the components of the special service circuit. Interest is applied to all these credits. As you can see interest payments can be quite an substantial portion of the refund check.

14 through 18 — Credit given for alll surcharges billed on the incorrect charges.

WHAT WILL HAPPEN IF I AM UNDERBILLED?

Don't panic. Most Bell companies won't try and make you pay back charges. They will start billing you from the day the error is discovered. It is only fair that you should pay for what you have. This can, of course, fall on the deaf ears of an unsympathetic boss. He could blame you for nosing around in the first place, rather that praising your desire to ensure proper telephone billing, which leads to our next category.

HOW TO DETERMINE IF YOU SHOULD USE A PROFESSIONAL BILL AUDITOR

How much time do you have to spend on the auditing process? Performing an audit can be a time consuming, frustrating process and does not always lead to a refund. How complex is your bill? Auditing the bill yourself allows you to keep all refunds and reductions in house. You may want to begin a review and see if you come up with any overbilling items. At that point you can decide if you feel comfortable with all the items on your CSR or if you should get help.

HOW TO SELECT AN AUDITOR

Selecting an auditor should not be any different from selecting a lawyer or other professional. Are they trustworthy? You are going to empower them to look over your billing records. What are the backgrounds of the auditors? You do not want amateurs reviewing your records. Get references. Call at least three of the references.

When interviewing auditors, question if they require you to make copies of all your bills. This can be quite time consuming. The auditor should be required to make his own copies of what he needs. All copies should be made at your premises. (Never let an auditor take your originals with him).

There is also a subjective quality to selecting an auditing firm. Do you feel comfortable dealing with the auditor? Remember you are authorizing them to act on your behalf. You do not want someone who is loud and obnoxious causing trouble in your company's name. What if there is a dispute between the auditor and yourself as to who found an overcharge? Do you feel that you can rationally resolve differences with the auditor? Interview at least three auditing companies to determine which one you feel most comfortable with.

Auditors vary as to what percentage of the refund they will get as payment. A typical agreement allows the auditor 50% of the refund that he gets from the telephone company. The agreement usually calls for the auditor to get a portion of future savings. For example, assume an auditor discovers that a circuit you are paying for was disconnected 5 years ago and gets you a $12,000 refund. The auditor will bill you $6,000. However, if that circuit currently costs $160 monthly, your next bill will be reduced by $160. A typical auditor will bill you 50% of the $160 monthly fee ($80) for 12 months.

You should not be billed by an auditor until you have received your refund or credit from the telephone company. You should also receive a full statement of all the credits obtained for you. If you have any doubts about any of the credits, have the telephone company provide a written letter stating what each credit is for, and who initiated the claim.

Do not select an auditor strictly on his contingency fee percentage. The lower the percentage the greater the probability that the auditor will only look for the "big kill" and will not thoroughly check your bill for all errors. Why? Because the time involved in finding relatively small errors won't be worth it. Also beware of people doing "free" audits or audits for extremely low contingency fees. Many are salesman in disguise and are looking to sell you services. Auditing provides them with a "foot in the door." They actually make their living selling local and long distance services and receive commissions on what they sell.

There is nothing wrong with an auditor making suggestions that will save you money, but the auditor ought to be very clear on whether or not he will receive a commission from another company if you agree to any of his recommendations.

Tariffs

Tariffs are documents filed by telephone companies that itemize prices for each of the services they offer. Local telephone company tariffs are approved by state commissions, while inter LATA service tariffs are approved by the FCC.

Before tariffs are accepted, hearings are held to determine if they are justified, and what impact they will have on subscribers. Only the major telephone companies require tariff approval.

In order to check if you are paying the correct rate for a particular service, you may need to look up the service in the telephone company's tariff. Tariff filings generally have a table of contents, so you can find out where the service you are questioning is listed. The tariff will describe the service, list pricing arrangements and any exceptions.

Tariffs contain thousands of pages and are constantly being changed and updated. You can subscribe to services that will send you tariffs for telephone companies throughout the United States. These services can be quite expensive. AT&T, in contrast provides free access to its tariffs for consultants. Even the call into the database is free (via an 800 number).

Most of the regulated local telephone companies make their tariffs available to the public. Most telephone books even have sections that list the tariffs for common telephone services.

It may be difficult, at first to find out where a company keeps its tariffs. Most telephone representatives will tell you that telephone company tariffs are kept on file at the public library. It is not true. Most libraries do not have these tariffs. An inquiry to the local Public Utility Commission will help you locate a telephone company office that makes tariffs available.

One word of advice. If you are researching tariffs in New York City, avoid the Public Service Commission's offices at 400 Broome Street. The office is dirty and the tariffs are rarely updated. Try New York Telephone's Public Office located at 204 2nd Avenue.

REDUCING YOUR TELEPHONE BILL THROUGH APPLICATIONS ENGINEERING

Applications engineering is the process of analyzing your telephone network to find products and services that will reduce your monthly bill without sacrificing network quality. It can be as simple as calling the telephone company to convert a particular service to a Rate Stabilization Plan (RSP). In many instances, the use of applications engineering concepts will increase the quality of your network. For example, putting DIDs onto a T1 will save you money and provide your network with a digital backbone. Unfortunately, most applications engineering is done by the telephone company or by their sales agents. Their main goal is not to save you money, but rather to sell telephone company products. Therefore, they are unlikely to advise you of all the hidden costs of converting to a particular service.

A true application engineer will provide you with a complete cost analysis that includes all the conversion costs, and provides you with the "break-even date." The break-even date is the date that your monthly saving offsets the initial conversion cost of the service. It is often used synonymously with the term break-even point.

CONVERTING DIDs TO A T1

Example #1-Replacing DID trunks with a T1 will save you money. The cutoff point is usually about 16 DIDs to a T1, depending on which telephone company is involved. To convert to a T1, you must be served by a digital CO. New York Telephone refers to this type of T1 as a FLEXPATH T1. Figure 3.10 compares a T1 that is connected to the CO to a point to point T1.

FIGURE 3.10

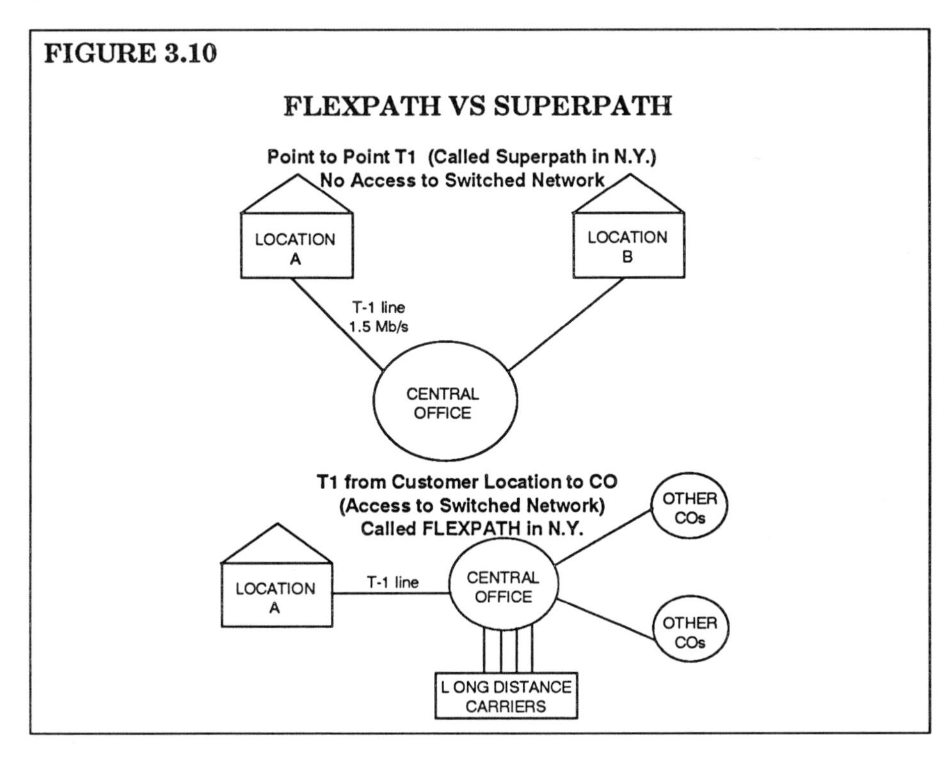

The following provides the break-even point for a conversion of 24 DID trunks to a T1 in New York City.

Step 1 — Price the full cost of a DID trunk. Charges for a DID usually include a trunk charge, loop charge and an FCC line charge. Check your CSR to get the exact rate that you are paying.

TB2 (DID Trunk Charge)	$56.17
D1F2X (loop charge)	$19.08
9ZR (FCC line charge)	$ 5.48
TOTAL COST OF A DID	$80.73

Step 2 — Determine the monthly cost of 24 DIDs. (A T1 to the CO allows you 24 ports or channels. You can fill these ports with any combination of 24 POTS or DID lines). The monthly cost of 24 DID lines is $1,937.52 (24 times $80.73).

Step 3 — Determine the monthly cost of a T1. To calculate the savings you will realize if you convert your DIDs to a T1, you need to know the distance between the customers' premises and his local CO. In the following example, the customer is located 1/2 mile from the CO.

You must include the cost of installing a T1 in your break-even point calculation. You can look up the tariff rates for a T1 in your area, or you can have the local telephone company provide you with a written estimate of the monthly costs for a T1 to the CO.

A FLEXPATH T1 from New York Telephone will cost you the following monthly and installation charges:

	Monthly	Installation
Service Order		$56.00
Premise Visit		$19.00
Group of 24 ports	$533.06	
Digital Transport Facility	$419.83	$1,550.00
24 FCC line charges @ $5.48	$131.52	
Total	$1,084.41	$1,625.00

Step 4 — Determine your monthly savings. Subtract the cost of a T1 from the monthly cost of 24 DID trunks. $1,937.52 minus 1,084.41 equals $853.11.

Step 5 — Determine the one time installation fee charged by your local telephone company for a T1. You can simply call the telephone company and have them provide you with installation costs. In this case, installation charges will be $1625.

Step 6 — Determine other conversion and installation costs. To utilize the T1, you will need a CSU (Channel Service Unit). A CSU regenerates digital signals and monitors your T1 for problems. It allows the local telephone company to test the circuit, should a problem arise. In addition, your PBX will probably need a T1 card to connect the T1 to the PBX. Assume this customer has a NEC 2400 IMS PBX. The following costs will apply:

A) T1 Card	$4,837.00
B) CSU	$3,083.00
Total	$7,920.00

Step 7 — Add the additional PBX installation costs to New York Telephone's installation costs to determine the total cost to convert to a T1. $1,625 plus $7,920 equals $9,545.

Step 8 — Calculate the break-even point by dividing the installation costs ($9,545) by the monthly savings ($853.11). Your break-even point is 11.19 months or just under a year. The monthly savings a T1 offers will pay for the installation costs within a year.

Savings over five years would be $51,186.60 ($853.11 x 60 months) minus the initial installation cost of $9,545.00 or $41,641.60. You should also know that by adding a T1 card you are also increasing the resale value of your PBX.

CONVERTING A COMBINATION OF POTS AND DID TRUNKS TO A T1

Example #2 — A customer located in New York City has 16 DID lines and 8 POTS trunk lines. Does it make sense for him to convert to a T1? Remember, a T1 gives you the use of 24 separate channels to the CO. You can combine DID, POTS and special service circuits onto one T1.

Step 1 — Determine your current monthly charges. A DID is $80.73 monthly and a POTS trunk line is $26.58 (1MB-16.23, TTB-4.87 & 9ZR 5.48).

16 DID Trunks	$1,291.68
8 CO Lines	$ 212.64
Total Monthly	$1,504.32

Step 2 — Determine monthly and installation cost from the telephone company for the T1. Assume that this customer is also within 1/2 mile of his CO. His installation and monthly costs will be the same as in the previous example.

T1 Monthly cost	$1,084.41
T1 Installation cost	$1,625.00

Step 3 — Determine your monthly savings:

Current monthly cost	$1,504.32
T1 monthly cost	$1,084.41
Monthly savings	$419.91

Step 4 — Total all installation costs. Assume that this customer also has a NEC 2400 IMS PBX. As previously determined, PBX installation costs will total $7,920. Total installation cost will be $9,545 ($1,625 plus $7,920).

Step 5 — Determine the break-even point. Divide your total installation cost ($9,545) by your monthly savings ($419.91) to get a break-even point of 22.73 months. This customer must keep the T1 a minimum of 23 months to justify the installation costs and to actually begin saving money.

RATE STABILIZATION PLANS (RSPs)

Sign a long term commitment to keep a particular telephone company service and they will usually give you a cheaper monthly rate. These plans are called different names by different telephone companies. They only make sense if you are sure that you will stay at the same location for an extended period of time.

Before you sign up for one of these plans, you should apply the principles of applications engineering to determine your true break-even cost. Many of these plans contain penalties if you should disconnect service before the commitment period expires. Many telephone company salespeople will focus on the monthly savings, and down play conversion costs.

TELEPHONE COMPANY SALES AGENTS

Many of the Bell Companies also sell their products through independent companies. These companies receive a commission based on the volume of products sold. They are highly motivated and they are very aggressive. They can also bundle telephone company products with other hardware and software offerings. This allows them to provide unique solutions to many problems.

The buyer of their services should be aware that these companies will receive a higher commission from the local telephone company if you sign long term contracts for products (RSPs). The longer the RSP (3 year, 5 year, 7 year), the greater the commission paid to them. The telephone company allows these independent companies to place the Bell logo on their cards and some of these companies take unfair advantage of this endorsement to confuse the customer. If you are approached by a unfamiliar telephone company sales agent, be aware that it is in your best interests to independently verify all cost savings.

POINT TO POINT T1 RSP

Example #3 — The following will show you how to determine the actual savings realized by a New York City customer that converts a T1 to a telephone company RSP:

Step 1 — From your CSR determine the current monthly cost for your T1. The following is how a point to point T1 will appear on a New York Telephone CSR:

Quantity	Item	Description	Rate
1	XUW1X	/CLS 96, DHDA, 999064,,N.Y. Mileage Charge	$129.78
2	3LN15	/CLS 96, DHDA, 999064,,N.Y. Mileage Charge	$ 83.06
	CKL	1-27 Broadway Manhattan, N.Y. /CLS 96, DHDA, 999064,,N.Y. /LSO 212 555/TAR 002	
	SN	CUSTOMER No. 1	
1	XUN1X	/EAR (Mileage Charge)	$269.93
2	XUD1Y	(Mileage Charge)	$ 62.30
	CKL	2-9 W42 Manhattan, NY /CLS 96, DHDA, 999064,, N.Y./LSO 212 999/TAR 002	
	SN	CUSTOMER No. 2 (BRANCH LOCATION)	
1	XUN1X	(Mileage Charge)	$269.93
3	XUD1Y	(Mileage Charge)	$ 93.45

The monthly charge for a this T1 can be summarized as follows.

Local Distribution Channel (between CO and customers' location CKL-1)		
	A) Fixed Charge (XUN1X)	$269.93
	B) Variable-31.15 per 1/4 mile (XUD1Y) 1/2 mile to CO	$ 62.30
Interoffice mileage between COs		
	A) Fixed (XUW1X)	$129.78
	B) Variable-41.53 per mile (3LN1S)-CO's are 2 miles apart	$ 83.06
Local Distribution Channel (between CO and customer's location CKL-2)		
	A) Fixed Charge (XUN1X)	$269.93
	B) 31.15 per quarter mile (XUD1Y) 3/4 mile to CO	$ 93.45
TOTAL		$908.45

Step 2 — Get the monthly rates for each RSP plan from the telephone company.

3 YEAR RSP

Local Distribution Channel (Location 1)		
	A) Fixed Charge (XUN1X)	$242.57
	B) Variable -$28.03 per 1/4 mile (XUN1X) 1/2 mile total	$ 56.06
Interoffice Mileage :Between CO's		
	A) Fixed (XUW1X)	$116.80
	B) Variable-$37.38 per mile (3LN1S) 2 miles total	$ 74.76
Local Distribution Channel (Location 2)		
	A) Fixed (XUN1X)	$242.57
	B) Variable-$28.03 per 1/4 mile (XUD1Y) -3/4 mile total	$ 84.09
TOTAL		$816.85

5 YEAR RSP

Local Distribution Channel (Location 1)		
	A) Fixed Charge (XUN1X)	$215.62
	B) Variable-$24.92 per 1/4 mile (XUD1Y) 1/2 mile total	$ 49.84
Interoffice Mileage Between CO's		
	A) Fixed (XUW1X)	$103.82
	B) Variable-$33.22 per mile (3LN1S) 2 miles total	$ 66.44
Local Distribution Channel (Location 2)		
	A) Fixed Charge (XUN1X)	$215.62
	B) $24.92 per 1/4 mile (XUD1Y) 3/4 mile total	$74.76
TOTAL		$726.10

7 YEAR RSP

Local Distribution Channel (Location 1)		
	A) Fixed Charge (XUN1X)	$188.66
	B) Variable-$21.80 per 1/4 mile (XUD1Y) 1/2 mile total	$ 43.60
Interoffice Mileage Between CO's		
	A) Fixed (XUW1X)	$ 90.84
	B) Variable-$29.07 per mile (3LN1S)-2 miles total	$ 58.14
Local Distribution Channel (Location 2)		

A) Fixed Charge (XUN1X) $188.66
B) Variable-$21.80 per 1/4 mile (XUD1Y)-3/4 mile total $ 65.40

TOTAL $635.30

Step 3 — Determine the monthly and yearly savings. In this example it is quite easy. Simply subtract the RSP monthly charges from your current charges.

This chart details the savings you will realize by switching to a RSP plan:

	Monthly Charge	**Yearly Charge**	**Yearly Savings**	**Total Savings Over RSP**
Current	$908.45	$10,901.40	—	—
3 Year Plan	$816.45	$9,797.40	$1,104.00	$3,312.00
5 Year Plan	$726.10	$8,713.20	$2,188.20	$10,941.0
7 Year Plan	$635.30	$7,623.60	$3,277.80	$22,944.6

Step 4 — Determine the break-even point. The cost to convert to a RSP is only $35.00 (a Record Order Charge). A seven year RSP will save your company $22,944.60 over the life of the contract.

Many companies have many T1s and pay month to month rates year after year. They are not aware of the savings available to them.

CENTREX:

Centrex can be used in place of a PBX. The CO acts like a PBX, and allows you all the features inherent in a PBX. Centrex can also be used in conjunction with a PBX (called the assume dial 9 option).

Many of the local telephone companies have been pushing rate stabilization plans for Centrex by paying high commissions to outside sales agents for signing long term Centrex customers.

One reason for this push is that competition is on the horizon. Companies like Teleport and MFS are starting to offer Centrex like services. While they are currently limited in scope, they expect to expand in the future. The goal of the local telephone companies is to lock in as many customers as possible. Before you consider a RSP for Centrex, call the Teleport and find out if they offer service in your area.If service from them is not currently available, they can tell when they plan to offer service in your area.

Example #4 — This customer is a branch office of a major New York City bank and has nine Centrex lines. New York Telephone charges a cheaper monthly rate for Centrex if you are served by an analog CO.

This customer has been at the same location for five years and has just signed a lease for another five years. He is served by an analog CO and is paying the telephone company month to month Centrex rates.

It is common for small branch offices of large companies (i.e. banks and insurance companies) to find themselves paying more for local telephone service than is necessary. Their central MIS or telecommunications department is often too busy to provide basic telephone direction to the branch offices. These central departments will concentrate on high technology offerings and leave the branch offices on their own with regards to their local telephone bill.

The following details cost savings involved if this customer converts to an Centrex RSP.

Step 1 — Check your CSR for your current cost for Centrex service. Then, call your local telephone company and find out the discount they will provide to you if you sign a long term contract.

When you check your CSR for the monthly cost of Centrex service, you will find that the billing differs from standard POTS billing. On the CSR you will notice that each Centrex line is listed as follows:

Quantity	Item	Description	Activity Date
1	RXR	CX 9410/HTG 100/LTC A3N/TBE B/PIC MCI/CTX 006	11/5/90
1	TDN	CX 9410 (Touch Tone Line)	3/21/90

Rates do not appear by the USOCs RXR and TDN as they would with POTS. Instead the charges are summarized at the end of the CSR as follows:

Quantity	Item	Description	Rate
9	9ZC	FCC Line Charge	$49.32
9	CIC	Primary Lines	$174.69
9	CEAC		$14.49
1	CYS	Common Equipment	$19.78

Every Centrex customer is billed one CYS (common equipment change) per account whether you have 9 lines or 99 lines.

The FCC line charge — 9ZC- (C for Centrex) also applies per line. The RXR charge is listed as CIC at the end of your CSR. The CIC charge is roughly the equivalent to the 1MB charge on a POTS line, and the CEAC charge replaces the touchtone charge.

Step 2 — Determine the monthly cost per item for a RSP from your telephone company. The longer the contract the greater the discount.

Cost Comparison of Centrex RSP Offerings

	Month to Month	1 Year RSP	2 Year RSP	3 Year RSP
CIC	$19.41	$15.26	$14.99	$14.72
CEAC	$ 1.61	$1.61	$1.61	$1.61
9ZC	$ 5.48	$5.48	$5.48	$5.48
CYS	$19.78	$12.11	$12.11	$12.11

Step 3 — Determine your current monthly charge for nine Centrex lines. In this example we will include the additional cost of surcharges levied on New York customers.

MONTH TO MONTH

(9) Centrex Exchange Access (CEAC)	$ 14.49
(1) Common Equipment (CYS)	$ 19.11
(9) Centrex Lines (CIC)	$174.69
(9) FCC Line Surcharge (9ZC)	$ 49.32
NY FCC Surcharge (.0132%)	$ 2.76
Municipal Surcharge (2.234%)	$ 4.67
N.Y. State/MTA Surcharge (5.84%)	$ 12.20
TOTAL	$277.24

Step 4 — Determine the monthly charge for each of the RSP plans.

ONE YEAR RSP	
(9) Centrex Exchange Access (CEAC)	$ 14.49
(1) Common Equipment (CYS)	$12.11
(9) Centrex Lines (CIC)	$137.34
(9) FCC Line Surcharge (9ZC)	$49.32
NY FCC Surcharge (.0132%)	$ 2.16
Municipal Surcharge (2.234%)	$ 3.66
N.Y. State/MTA Surcharge (5.84%)	$ 9.57
TOTAL	$228.65

TWO YEAR RSP	
(9) Centrex Exchange Access (CEAC)	$ 14.49
(1) Common Equipment (CYS)	$ 12.11
(9) Centrex Lines (CIC)	$134.28
(9) FCC Line Surcharge (9ZC)	$ 49.32
NY FCC Surcharge (.0132%)	$ 2.12
Municipal Surcharge (2.234%)	$ 3.59
N.Y. State/MTA Surcharge (5.84%)	$ 9.40
TOTAL	$225.31

THREE YEAR RSP	
(9) Centrex Exchange Access (CEAC)	$ 14.49
(1) Common Equipment (CYS)	$ 12.11
(9) Centrex Lines (CIC)	$132.48
(9) FCC Line Surcharge (9ZC)	$ 49.32
NY FCC Surcharge (.0132%)	$ 2.10
Municipal Surcharge (2.234%)	$ 3.55
N.Y. State/MTA Surcharge (5.84%)	$ 9.29
TOTAL	$223.34

Step 5 — Chart the savings for each RSP plan.

	Monthly Charge	**Yearly Charge**	**Yearly Savings**	**% Yearly Savings**	**Total Savings Over RSP**
Current	$277.91	$3,334.92	None	0%	0
1 Year RSP	$228.65	$2,743.80	$591.12	17%	$591.12
2 Year RSP	$225.31	$2,703.72	$631.20	19%	$1,262.40
3 Year RSP	$223.34	$2,680.08	$654.84	24%	$1,964.52

Step 6 — Determine the cost to convert to an RSP.

You can call the telephone company and request the conversion charges. The charge to convert to an RSP in New York is one Record Order Charge (ROC) of $35.00. That is the total cost regardless of the number of lines you have.

Step 7 — Determine your break-even point by dividing your conversion costs by the monthly savings. Your first months savings pays for conversion cost (ROC charge of $35.00). Your break-even point is one month regardless of the plan you chose. A 3 year RSP will save you a total of $1,964.52. Not bad for 9 lines !

POTS VERSUS ANALOG CENTREX

Some Bell Companies sell analog Centrex at a price than is lower than the cost of a POTS line, while digital Centrex is sold at a more expensive price. The goal is to entice customers to convert to a long term Centrex contract by offering a better rate. There are, however, significant conversion costs involved in switching from POTS to Centrex service. The advantage to the telephone company is that, in addition to the revenue generated by one time conversion costs, they also lock you into what is actually a premium service.

When the telephone company finally gets around to upgrading your CO to a digital switch (and after your contract expires), you will automatically get billed the higher digital Centrex rate. If you decide to convert back to POTS at a later date, you will incur additional conversion costs.

Analog Centrex is usually marketed to small customers who are told they will save money. However, unlike switching from an existing Centrex billing plan to a RSP, significant conversion costs apply. Depending on the number of lines and term plan you sign for, it may not be cost justified to switch. The following example details how you can determine the break-even point for converting a POTS line to an analog Centrex RSP.

Example #5

Step 1 — Determine from your CSR what you currently pay for POTS service. In this example, Customer A has 12 POTS lines, while Customer B has 12 PBX POTS trunk lines (some telephone companies charge more for PBX trunk lines). For illustrative purposes we will exclude surcharges.

Customer A

1 x 16.23 (1MB)	$ 16.23
11 x 16.23 (ALN)	$178.53
12 x 3.08 (TTB)	$ 36.96
12 x 5.48 (9ZR)	$ 65.76
TOTAL	$297.48

Customer B has 12 PBX trunk lines.

1 x 16.23 (XMB)	$ 16.23
11 x 16.23 (TXM)	$178.53
12 x 4.87 (TJB)	$ 58.44
12 x 5.48 (9ZR)	$ 65.76
TOTAL	$318.96

Step 2 — Call the telephone company and get the price for a 3 Year RSP on analog Centrex service.

12 x 14.72 (CIC)	$176.64
12 x 1.61 (CEAC)	$ 19.32
12 x 5.48 (9ZC)	$ 65.76
1 x 12.11 (CYS)	$ 12.11
TOTAL	$273.83

Step 3 — Calculate your monthly savings. The following chart details the cost comparison between a 3 year RSP for analog Centrex, POTS and PBX POTS trunk line service for a customer with 12 telephone lines.

	Monthly Charge	Yearly Charge
Centrex (3 Year RSP)	$273.83	$3,285.96
POTS	$297.48	$3,569.76
PBX Trunk	$318.96	$3,827.52

The monthly savings need to be cost justified by calculating the full installation price and determining your break-even point.

Step 4 — Determine all conversion costs. The following details the conversion cost for 12 telephone lines. One time conversion costs are the same for both 1MB and trunk lines.

SERVICE CHARGE	$ 56.00
REWIRE CHARGE(PER LINE) $26.05 x 12	$312.60
COMMON EQUIPMENT CHARGE	$ 59.36
TOTAL	$427.96

Step 5 — Determine your break-even point. The one time charge to convert to analog Centrex is $427.96. To calculate the break-even point for POTS service, take the monthly savings of $23.65 ($297.48 - $273.83) and divide it into your conversion cost of $427.96. The break-even cost is 18.1 months.

The break-even point for a POTS trunk customer is calculated as follows:

Monthly savings equals $45.13 ($318.96 - $273.83). Divide the conversion cost ($427.96) by $45.13 to get a break-even point of 9.48 months.

INTER-LATA CORRIDOR SERVICE

In most areas, the local Bell Telephone Companies are prohibited from providing inter-LATA service. Two exceptions to this rule is what is called the New York/New Jersey Corridor and the Camden/Philadelphia Corridor. Both of these cities are considered so economically intertwined that the FCC allows the local telephone companies to provide service between these areas even though LATA boundaries are crossed.

In the New York/New Jersey Corridor, both New York Telephone and New Jersey Bell can provide service. (See figure 3.11 on the following page).

You can get dedicated lines across these boundaries from either your local telephone company or from a long distance carrier. Services obtained from a long distance carrier will be up to three times more expensive than those provided by New York Telephone and New Jersey Bell. Many companies end up paying more for services within the corridor simply because they are not aware of their options.

Our next example illustrates the difference in billing for two voice grade datalines located within the New York/New Jersey Corridor. It illustrates the difference in pricing between a circuit provided by New York Telephone and New Jersey Bell as opposed to one provided by a long distance carrier.

When you order inter-LATA services from a long distance

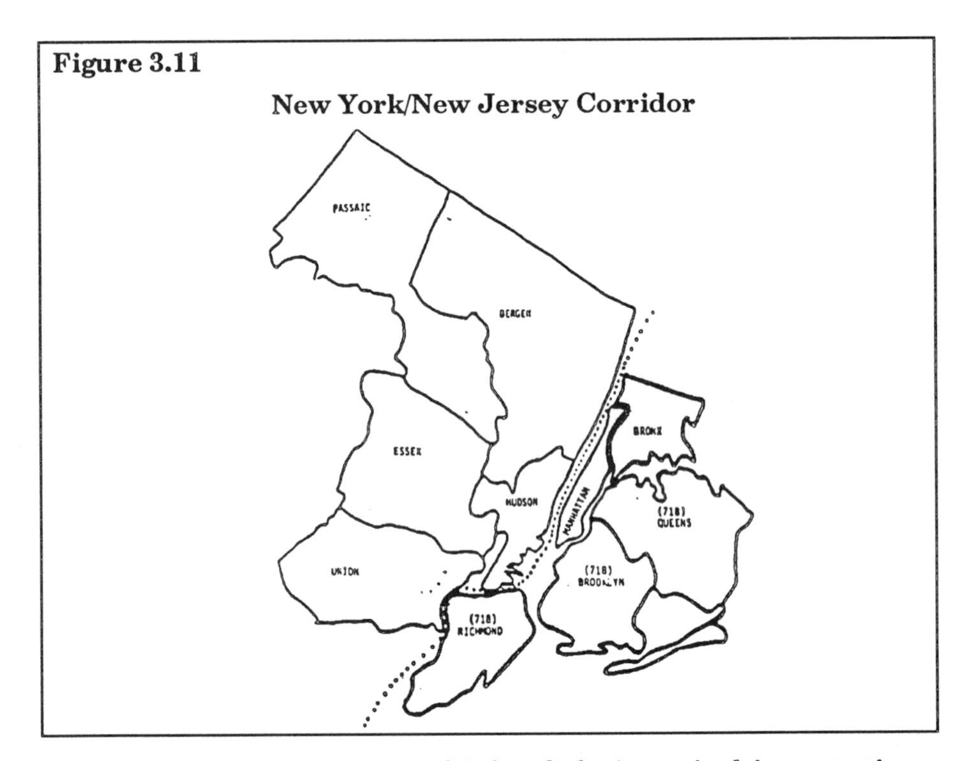

Figure 3.11

New York/New Jersey Corridor

carrier, they determine which of their switching stations or Points of Presence (POPs) are closest to the locations you wish to connect. They then provision a circuit that connects the POPs. The long distance carrier will then contract out the portion of the circuit that connects their POP to the customer premises. This portion of the circuit runs within a LATA and is usually supplied by the local telephone company.

The local telephone company actually connects the circuit from the long distance carrier's POP to your local telephone CO, and then to your location. The local telephone company bills the long distance carrier for the circuit based on the distance (mileage) from the POP to the local CO. The long distance carrier provides the end user with one bill that covers the cost of connecting their POPs, and that also covers the cost billed by the local telephone company.

The long distance carrier "marks up" the cost charged to them by the local telephone company. They also add an access coordination fee for ordering the local portion of the circuits. Figure 3.12 illustrates a typical inter LATA connection.

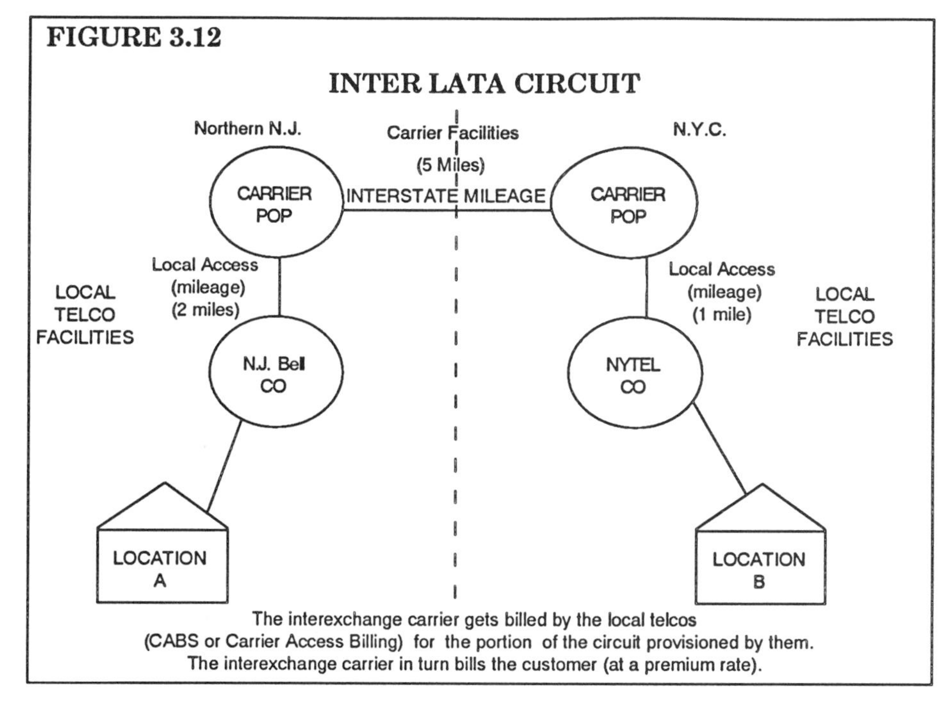

Example #6 — Customer has two offices that he wishes to connect via a private data line. Both ends of the circuit are located within the New York/New Jersey Corridor. Location #1 is located in North Bergen, New Jersey, and location #2 is located in Midtown Manhattan.

This company originally used Metromedia (ITT) to provide two inter LATA data lines. Metromedia's POPs in each LATA are five miles apart. The distance from the POP to the local CO is 2 miles at the New Jersey end and 1 mile at the New York end. (See figure 3.12)

Metromedia bills for this circuit as follows:

TELEPHONE BILL FROM METROMEDIA FOR ONE INTER LATA VOICE GRADE DATA CIRCUIT

5 miles (interstate mileage between POPs)	$152.41
1 mile (local access in New York)	$112.47
2 miles (local access in New Jersey)	$152.29

Central Office Connection (to connect the circuit to New York Telephone's CO)	$ 19.50
Central Office Connection (to connect the circuit to New Jersey Bell's CO)	$ 19.50
Access Coordination Fee (New York portion)	$ 10.50
Access Coordination Fee(New Jersey portion)	$ 10.50
TOTAL	$477.17

NOTE: The circuit travels a total of eight miles in this configuration. Two datalines will cost you $954.34 monthly.

Since this circuit is located in the New York/New Jersey Corridor you can "cut out the middleman" i.e. Metromedia, and obtain service directly from the local telephone companies.

The same exact service can be ordered from either New York Telephone or New Jersey Bell, at a much cheaper price. It does not matter which of the companies you contact first, as each one will coordinate the service with the other company.

Each dataline from the local company is configured as shown in figure 3.13 on the following page.

By bypassing Metromedia's POPs, your circuit's path becomes more direct. The distance between your local New Jersey Bell CO and your New York Telephone CO now becomes only 4 miles. When the circuit was ordered via Metromedia, the circuit travelled a total of eight miles.

BILLING OF CORRIDOR CIRCUIT

The local telephone companies split the billing for corridor circuits by means of the BIP identifier. The BIP identifier is a FID that stands for Billing Percentage. Figure 3.13 on the following page details how a corridor circuit is configured.

Mileage is billed by New York Telephone and New Jersey Bell according to what percentage of the circuit is physically located within each state. You will receive two separate bills from each company. The bills will be assigned a special bill number for tracking purposes. The following shows that the total of the bills from both local companies are only 33% of the cost from Metromedia for the same circuits.

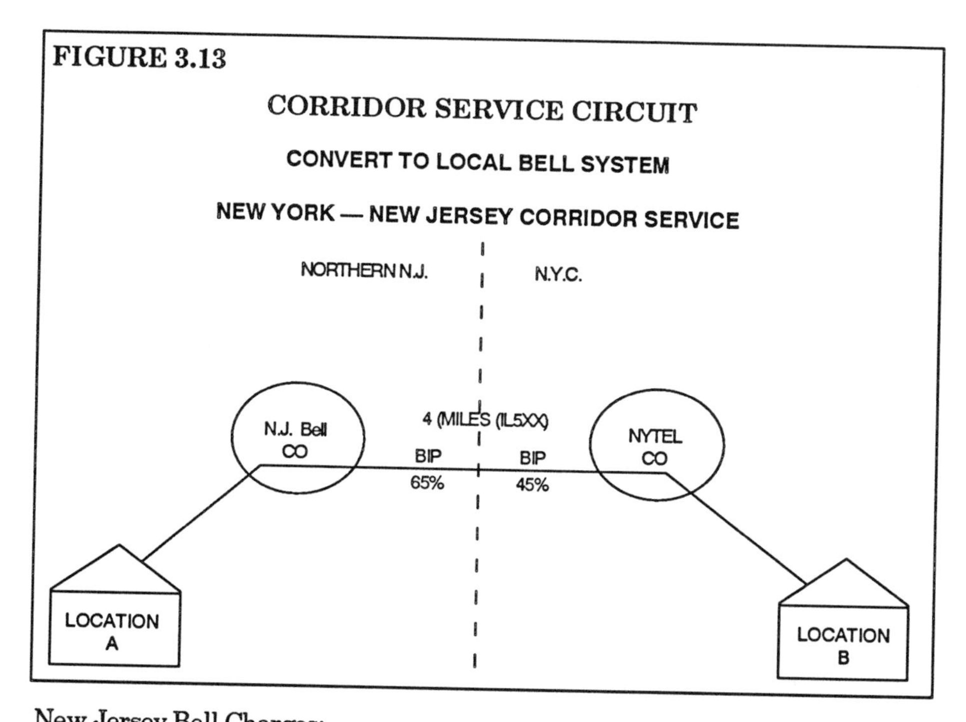

New Jersey Bell Charges:

Quantity	USOC	Description	Rate
1	T6E4X	Channel Termination	$41.36
4	1L5LS	Mileage(BIP=65%)	$11.83
TOTAL			$53.19

N.Y. Telephone Company Charges:

Quantity	USOC	Description	Rate
1	T6E4X	Channel Termination	$55.44
4	IL5XX	Mileage (BIP = 45%)	$21.24
TOTAL			$76.68

Total charges for this circuit totals $129.87 ($53.19 plus $76.68). The cost for two circuits will be $259.74. This customer will save money by cancelling his circuits with Metromedia and re-ordering them directly from New York Telephone and New Jersey Bell. To calculate the break-even point for this customer follow the following steps:

Step 1 — Calculate monthly savings (customer has two datalines) by subtracting the local telephone company's monthly charges from Metromedia's monthly charges.

Metromedia (2 times $477.17)	$954.34
Local Bell Companies (2 times 129.87)	$259.74
Monthly Savings	$694.60

Step 2 — Call the local telephone company and get the total installation charges for a corridor service circuit.

Installation Charges Per Circuit - New Jersey Bell

USOC	Description	Charge
T6E4X	Channel Termination	$637.58
Total for two circuits		$1,275.16

Installation Charges Per Circuit- New York Telephone

USOC	Description	Charge
T6E4X	Channel Termination	$375.68
Total for two circuits		$751.36

Total installation charges for both New York Telephone and New Jersey Bell equals $2,026.52.

Step 3 — Determine the break-even point:

Divide the total monthly savings ($694.60) by the total installation cost of $2,026.52. The break-even point is 2.92 months.

This change to corridor service will result in a savings over 5 year period will equal $41,676.00 minus $2,026.52 (installation cost) or $39,649.48. Not bad for converting two datalines to local Bell service.

INTER-LATA CSRs

Your monthly bill for a inter-LATA corridor circuit is rendered under what is known as a special bill number. New York Telephone's special bill format is M51-XXXX, while New Jersey Bell's format is R15-XXXX. To get an itemization of the charges that comprise your bill you have to order a CSR.

The CSR for a inter-LATA circuit is very similar to the CSRs for intra-LATA special circuits reviewed in chapter 2. The main difference is that the services (USOCs) are tariffed by the FCC, since they cross LATA and state boundaries. To help you read a corridor service CSR, the following provides definitions of common inter-LATA USOCs.

COMMON INTER-LATA USOCs:

ACNA	ACCESS CUSTOMER NAME ABBREVIATION
CKL	CIRCUIT LOCATION
CLS	CIRCUIT ID - SERIAL NUMBER FORMAT
LAT	LOCAL ACCESS TRANSPORT AREA ID
LOC	LOCATION
LSO	LOCAL SERVING OFFICE
MPB	MEET POINT BILLING
NC	NETWORK CHANNEL
NCI	NETWORK CHANNEL INTERFACE
PF	PRINT FREQUENCY
PIU	PERCENTAGE OF INTERSTATE USAGE
SN	SERVICE NAME
ZCR	CORRIDOR SERVICE
S25	SPECIAL ACCESS SURCHARGE
T6E2X	CHANNEL TERMINATION 2 WIRE
T6E4X	CHANNEL TERMINATION 4 WIRE
UTM	SPECIAL ACCESS RECOVERY SURCHARGE
XDV2X	ANALOG VOICEGRADE 2 WIRE
XDV4X	ANALOG VOICEGRADE 4 WIRE
XSSLR	20HZ RINGDOWN INTERFACE
1L5LS	MILEAGE-
1RL2W	VOICEGRADE IMPROVED RETURN 2 WIRE
1RL4W	VOICEGRADE IMPROVED RETURN 4 WIRE

A corridor Circuit ID will be in serial number format and starts with the prefix 32. Just as with intra-LATA special service circuits, the Circuit ID can be broken down into its components. A sample Circuit ID of 32,LGGS,999123,,NY would be divided as follows:

	PREFIX CODE	SERVICE	MODIFIER NUMBER	SERIAL
POSITION #	1 2	3 4	5 6	7 8 9 10 11 12
CIRCUIT ID	3 2	LG	GS	9 9 9 1 2 3

NOTE: THE NY TELLS YOU THAT THE CIRCUIT WAS ORIGINATED BY THE NEW YORK TELEPHONE COMPANY

The service code tells you for what purpose the circuit is used. Inter-LATA circuits have different service code definitions. The following table lists the definitions for common service codes.

INTER-LATA SERVICE CODES (Circuit Position #3 & #4).

SERVICE CODE	DEFINITION
HC	HIGH CAPACITY 1.544 Mbps
HD	HIGH CAPACITY 3.152 Mbps
HE	HIGH CAPACITY 6.312 Mbps
HF	HIGH CAPACITY 44.736 Mbps
HG	HIGH CAPACITY 274.176 Mbps
LB	VOICE-NON SWITCHED
LC	VOICE-SWITCHED LINE
LF	LOW SPEED DATA
LG	BASIC DATA LINE
LH	TIE TRUNK
LJ	VOICE AND DATA SSN ACCESS
LK	VOICE AND DATA-INTERMACHINE TRUNK
LN	DATA EXTENSION
SB	SWITCHED ACCESS-STANDARD
SD	SWITCHED ACCESS-IMPROVED
SE	SPECIAL ACCESS-DEDICATED
SF	SPECIAL ACCESS-DEDICATED IMPROVED
WB	WIDEBAND DIGITAL 19.2 kb/s
WE	WIDEBAND DIGITAL 50kb/s
WF	WIDEBAND DIGITAL 230.4 kb/s
WH	WIDEBAND DIGITAL 56 kb/s
XA	DEDICATED DIGITAL 2.4 kb/s
XB	DEDICATED DIGITAL 4.8 kb/s
XG	DEDICATED DIGITAL 9.6 kb/s
XH	DEDICATED DIGITAL 56 kb/s

A sample New York Telephone CSR is explained as follows:

1	QTY	CODE	DESCRIPTION	RATE
2		CLS	32,LNGY,999179,,NY	
3			/PIU 100	
4		CKL	1-270 MADISON AVE,	
5			MANHATTAN, NY/LSO 212555	
6			/SN ABC CONTEMPORARY.	
7	4	1L5XX	/45 BIP X $29.58 (4 x $4.41)	$21.24
8	1	T6E4X	CHANNEL TERMINATION	$55.44
9		CKL	2-15 BERGEN RD, NORTH	
10			BERGEN, NJ	
11			/SN ABC CONTEMPORARY	

Line 1— Lists the headings to the columns and is very similar to an intra-LATA CSR except that "code" is used instead of "item" to denote USOCs and quantity is abbreviated with QTY.

Line 2 — The USOC CLS tells us that the Circuit I.D. is in serial number format. The Circuit I.D. in this example is 32,LNGY,999179,,NY.

Line 3 — PIU 100 indicates that the service is used exclusively (100%) for interstate transmission

Line 4 — CKL 1 gives us the address of the originating location of the circuit.

Line 5 — The Local Serving Office (LSO) is 212 555.

Line 6 — The SN identifier tells us the name of the customer at the originating location.

Line 7 — The 1L of 1L5XX always indicates mileage charges will follow. Here, under QTY (Quantity), the 4 indicates four miles. Interstate corridor mileage is billed via a standard formula. A fixed minimum mileage cost of $29.58 is added to the variable mileage cost. The variable cost is determined by multiplying the total number of miles by $4.41 (4 times $4.41). The total mileage cost in this case is $21.24.

Since the total mileage charge is split between New York Telephone and New Jersey Bell we have to factor in the BIP (billing percentage). The BIP indicates what percentage of the circuit lies within the boundaries of New York as opposed to that portion that lies within New Jersey.

The BIP on your New York Telephone CSR plus the BIP on your New Jersey CSR should always equal 100%.

Line 8 — T6E4X is the USOC code for a 4 wire channel termination charge.

Lines 9, 10, 11 — CKL 2 is the USOC that denotes the location of the terminating end of the circuit. The SN USOC denotes the name of the company at the terminating end. In this case its a branch office of the same customer, ABC Contemporary.

New Jersey Bell will also send (upon request) you a CSR for its portion of the billing of a corridor circuit. This CSR is for the same circuit as detailed on the New York CSR. The CSRs are very similar. The following is an abbreviated CSR.

QTY	CODE	DESCRIPTION	AMOUNT
	CLS	LNGY 999179..NY / PIU 100	
	CKL	1-270 Madison Ave, (New York, N.Y.) LSO 212555	
	T6E4X	CHANNEL TERMINATION	$41.36
4	1L5LS	$15.00 + (4 X .80) x BIP 65	$11.83

New Jersey Bell charges different rates for fixed and variable mileage, although the formula to calculate total mileage is the same. New York Telephone bills a fixed minimum mileage charge of $33.39 while New Jersey Bell bills just $15.00 for the fixed mileage. The per mile variable rate is only 80 cents in New Jersey.

COMMON CORRIDOR SERVICE BILLING ERRORS

The single most common error on corridor service (and all interstate circuits) is the application of the Special Access Surcharge Exemption (S25 USOC).

FCC Tariff No. 40/41 Paragraph 5.7 (E)(2) allows the telephone companies to place an additional surcharge on circuits that access the switched network. This surcharge is automatically placed on your bill by the telephone company unless you file an exemption. A separate exemption must be filed in both New York and in New Jersey.

It is your responsibility to determine and inform the telephone company if you are exempt from this surcharge. Your circuit is exempt from the special access surcharge if it meets any of the following conditions:

1. An open end termination in a telephone company switch (FX line).

2. An analog channel termination that is used for radio or television program transmission.

3. A termination used for TELEX service.

4. A termination, that by the nature of its operating characteristics, could not make use of telephone company common lines.

5. A termination, that interconnects either directly or indirectly to the local exchange network, where the usage is subject to Carrier Common Line charges.

6. A termination, that the customer certifies to the telephone company is not a PBX or other device capable of interconnecting the private line facility to a local subscriber line.

In essence, the circuit should be exempt if it is used for point to point transmission and can not access the public switched network. That covers the majority of circuits.

On a CSR provided by New York Telephone, you can identify the surcharge as follows:

CODE	DESCRIPTION	AMOUNT
S25	INTERSTATE 100%	$25.00

On a New Jersey Bell CSR, the surcharge is comprised of two charges; the S25 and UTM USOCs.

CODE	DESCRIPTION	AMOUNT
S25	INTER NJ 100%	$25.00
UTM	INTER NJ 100%	$ 4.56

If you have an existing circuit and now realize it is surcharge exempt, notify the telephone company. New Jersey Bell and New York Telephone representatives will give you three months backcredit from the date they receive your exemption form.

An exception is granted if you have proof that an exemption certificate was sent in at an earlier date, but was not properly applied.

HOW TO IDENTIFY AND OBTAIN REFUNDS ON CORRIDOR CIRCUITS

The key to auditing corridor service circuits is to compare the New York Telephone CSR to the New Jersey Bell CSR. The rates may be different, but they should otherwise be mirror images of each other.

If, for example, your New York Telephone CSR does not have a S25 surcharge, but your New Jersey Bell CSR does have one, a bell should go off. Even if you do not have records to prove that you sent New Jersey Bell an exemption certificate you now have room to negotiate with them.

To better negotiate with the telephone company you should have an understanding of how New Jersey Corridor circuits are provisioned.

Step 1 — You call up the New York Telephone Company and request a voice grade data circuit between New York City and Northern New Jersey.

Step 2 — The New York Telephone representative takes your request, and enters your order into a software system which generates what is known as an Access Service Request (ASR).

There are national standards that detail the format of an ASR. The ASR contains a field that lets all standard telephone company billing systems know whether a Special Access Surcharge should be billed on a particular circuit. This field is appropriately called the S25 field. The representative must enter a Y (yes, charge the S25) or an N (no charge, the customer is exempt) in this field. The order is then sent to downstream New York Telephone systems to provision the New York portion of the circuit. The order is simultaneously sent to New Jersey Bell via a mechanized data link.

The order is down-loaded by New Jersey Bell and converted so it can be read by New Jersey's Bell's service order processor. If the order is transmitted by New York Telephone with a Y instead of an N in this field, New Jersey Bell will also bill the S25 surcharge.

When negotiating refunds with either of the Bell companies, you can point out how complicated this system for coordinating orders can be. It also affords numerous opportunities for errors. You should not be penalized for errors caused by either Bell company.

COMMON MILEAGE ERRORS

The second most common corridor service error is the incorrect calculation of mileage. As previously mentioned, New York Telephone and New Jersey Bell split mileage charges for a corridor circuit based on the portion of the circuit that lies in each state.

The BIP FID details the percentage of the circuit that lies within each state. The sum of the BIPs from New York Telephone and New Jersey Bell should never exceed 100%. The following example illustrates overbilling for mileage that is detected by adding the BIPs on the New York Telephone and New Jersey Bell CSRs. The Circuit ID is 32, LNGS,,999123. Compare the CSRs and add the BIPs together.

New Jersey Bell CSR

QTY	CODE	DESCRIPTION	AMOUNT
8	1L5XX	[$15.00+ (8 x .80) x 85 BIP]	$18.19

New York Telephone CSR

QTY	CODE	DESCRIPTION	AMOUNT
8	IL5XX	[$29.58+(8 x 4.41) x 45 BIP]	$29.18

When you add up the BIPs (85 + 45) they exceed 100%.

In this example you are overbilled and should get backcredit from the date the circuit was installed. To calculate your savings first determine what the correct BIPs should be. Substitute the new BIPs into the mileage formula detailed on your CSR. In this example, New York Telephone's BIP should be 15% instead of 45%.

Calculate the correct mileage charge by substituting the correct BIP into the mileage formula as follows: [$29.58 x (8 x 4.41) x 15 BIP]. The correct charge from New York Telephone should be $9.72. You have been paying New York Telephone Company $29.18 and you are due a backcredit of $19.31 ($29.18 minus $9.72) per month the circuit was in existence. You should be credited $19.31 per month back to the installation date (plus taxes, surcharges and interest). An interstate circuit credit is not given back more than two years due to the Communications Act of 1934.

CHAPTER 4:

LOCAL AND LONG DISTANCE CALLING

UNDERSTANDING YOUR LONG DISTANCE BILL

There are two ways to access the long distance network. The first way to gain access is through your local telephone company (switched access). In this method of access, your call travels over your existing telephone lines. Access is gained by using the local telephone company's switching facilities. Figure 4.1 on the following page details this method of access.

The local telephone company charges the long distance carrier for use of their facilities. Your long distance carrier includes this cost in their bill to you.

The second method of access is a via dedicated line (i.e. T1) that connects you directly to your long distance carrier's Point of Presence. (POP). This method of access is commonly called bypass because your call does not "pass through" the local telephone company's CO.

In this configuration, the customers PBX switch determines the path of the call. Local calls are forwarded to the local

FIGURE 4.1

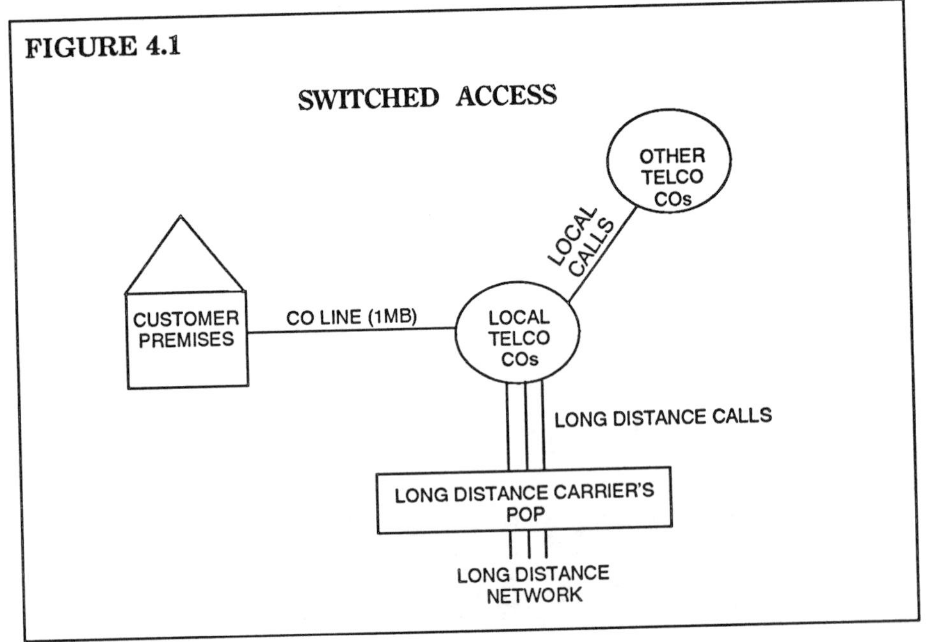

telephone companies' switch. Inter LATA calls are sent directly to the long distance carriers' POP, via a dedicated line. Figure 4.2 details this method of access.

All long distance calls made via the switched network are first processed by the local telephone company. When your local CO receives your call, it must first determine if the final destination of the call lies outside the LATA. If the call

FIGURE 4.2

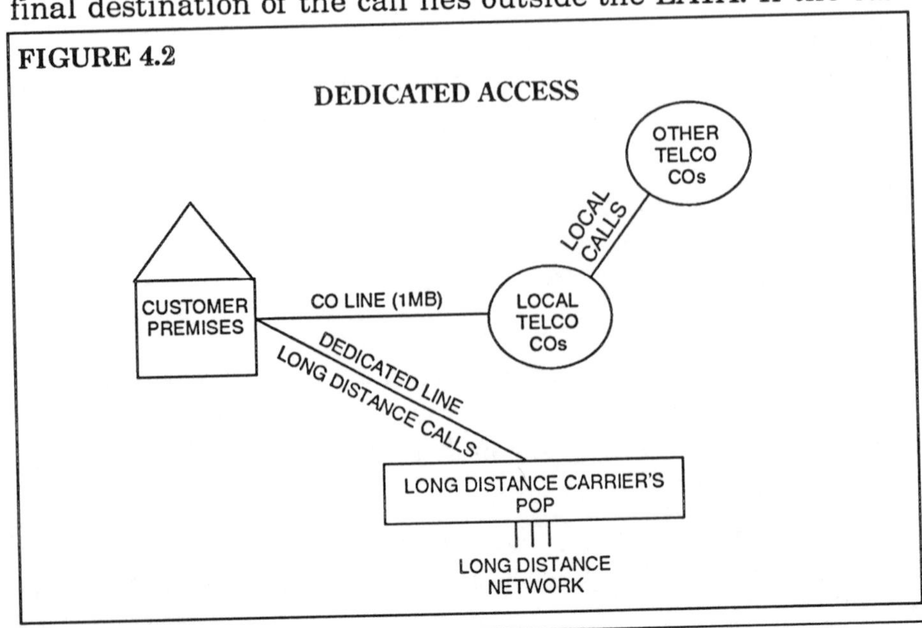

terminates outside the LATA, your PIC (primary carrier selection) is then determined and your call is sent to your long distance carrier of choice. The local company charges your long distance carrier for each call sent to them. Approximately 25% of the cost of a long distance call is paid by the long distance carrier to the local telephone company.

RESELLERS

Under most circumstances, AT&T charges the highest rates. There are, however, ways to get AT&T's service for less.

One such method is to sign up for AT&T's SDN (Software Defined Network) service.

SDN offers significant discounts to large users. In return for lower calling rates, you must guarantee that you will meet minimum usage requirements. Smaller companies can qualify for the lower SDN rates by utilizing the services of an SDN Reseller.

SDN Resellers bundle many small users together so they can meet minimum usage requirements. Resellers form a sort of unofficial association that qualifies small businesses for SDN discounts.

Resellers can also offer discounts off the rates of other long distance carriers. One example is National Communications Association's (NCA) TRT METROSELECT service. TRT service offered through NCA is actually cheaper for smaller companies than if they obtained service directly from TRT.

When you use a Reseller, your bill comes directly from the carrier they are associated with. For example, your SDN bill will come directly from AT&T. Rebillers, on the other hand, obtain the billing tapes from the carriers they are associated with. They in return re-rate the calls according to their own price structure.

Selecting a Long Distance Carrier

To get the best long distance rates, follow the following guidelines:

1. First, and most importantly, place your long distance business out to bid. Contact AT&T, MCI, TRT, SPRINT, CABLE & WIRELESS, along with a Reseller to prepare a bid for your service.

As a result, you will get a written statement of your charges and projected savings from each carrier. Have each carrier point out the "cons" of their competitors' bids. You will receive quite an education from this process. You will also have a statement from which to verify your actual savings.

2. Avoid signing contracts that commit your company's long distance to a particular carrier for more than one year. With rates and promotions in a state of constant flux, you want to have the flexibility to take advantage of market changes.

The cost of a long distance call is based on the distance the calls travels. The terminating point of your call is assigned to a range or mileage band. The call is rated by the long distance carrier based on the band in which your call terminates. Mileage Bands are grouped as follows:

BAND 1	0-55 miles
BAND 2	56-292 miles
BAND 3	293-430 miles
BAND 4	431- 925 miles
BAND 5	926-1910 miles
BAND 6	1911-3000 miles

FIGURE 4.3

SAMPLE LONG DISTANCE RATES

MILEAGE BAND	AT&T PROWATS	MCI PRISM PLUS	C&W FOCUS III	SPRINT DIAL 1	AT&T SDN	NCA/TRT METRO-SELECT
0-55	.220	.199	.184	.181	.179	.139
56-292	.240	.223	.205	.197	.198	.154
293-430	.254	.239	.208	.216	.211	.160
431-925	.254	.247	.225	.222	.219	.160
926-1910	.265	.257	.227	.231	.231	.160
1911-3000	.265	.257	.233	.231	.231	.160
AVERAGE RATE PER MINUTE	.249	.237	.214	.213	.213	.156

Figure 4.3 on page 130 details the price per band charged by carriers for typical long distance services.

LONG DISTANCE RATE COMPARISON

(Switched Banded Service-Domestic Day Interstate Rates)

If you install a T1, your long distance bill will drop about 25%. If, for example, you are on MCI's VISION service, you will pay 21 cents for calls made over the switched network. With T1 access to MCI's POP, you will only pay 15.5 cents per minute for the same call.

Rate comparisons can be misleading. For most services, AT&T will be the most expensive carrier followed by MCI. Each carrier, however, offers a variety of services and options, which can make it difficult to accurately compare rates.

Many carriers will reduce their tariffed rates to make a sale. Keep in mind that the smaller, less known carriers like TRT have lower overhead and therefore have more flexibility in pricing. Finally, before you settle on a long distance carrier ask yourself how much of your money will be applied to the carrier's advertising budget, and if you really get better quality from AT&T.

LONG DISTANCE CALLING AND THE ECONOMICS OF A T1

As previously noted, installing a T1 will reduce your long distance bills by approximately 25%. So why doesn't everyone install a T1? You need to have a certain volume of traffic to justify a T1. To calculate your breakeven point, you need to use the principles of applications engineering. First you must factor in the total cost of installing a T1, and the monthly rental of a T1. As detailed in chapter 2, the cost of a T1 depends on the distance your local CO is to the long distance carrier's POP. The following determines the savings a customer with a monthly long distance of $8,000 will realize by installing a T1.

Example #1 — Assume total installation cost for your T1 (includes T1 card, CSU and install charge) is $9,545, monthly rental on the T1 is $950 and your average monthly long distance bill is $8,000. A 25% rate reduction is quoted by your long distance carrier if you install a T1.

Step 1. — Determine your monthly savings on your long distance bill. $8,000 x 25% = $2,000

Step 2. — Determine actual monthly savings. Subtract monthly savings in long distance usage from the fixed monthly cost of the T1. $2,000 - $950 =$1,050

Step 3. — Determine your breakeven point: Divide the total installation cost by the actual monthly savings. $9,545 divided by $1,050 = 9.09 months. The T1 pays for its installation in 9.09 months. From that point on the customer realizes a monthly savings of $1,050 or a 13% actual reduction on a $8,000 bill.

CALCULATING LONG DISTANCE CHARGES

Before you can audit your current long distance bill you need to understand how your long distance carrier calculates the charge for a call.

To price domestic calls you need to know the carrier's band rates. For international calls you need to know the rate for the initial period (usually the first 30 seconds) and then the rate for each additional interval (usually six second intervals) for the country dialed. (These rates should have been documented as part of the bidding process. If you did not put your long distance out to bid, call your current carrier directly and get these rates in writing).

To calculate domestic calls:

Step 1 — Determine the band in which the call terminates.

Step 2 — Take the length of the call and multiply it by the band rate per minute.

Example: Assume a Band 1 rate of .164 per minute. Call is for 2.4 minutes.

2.4 minutes x .164 = .3936 or 39 cents. The cost of your call should be 39 cents.

To calculate the price of an international call follow the following steps:.

Step 1. — Determine the cost of the first minute.

30 second initial price + 6 second price x 5 = 1st minute cost

Step 2 — Calculate Additional Minute Cost:

6 second price x additional 6 second increments = Additional minute cost.

Step 3 — Add first and additional minutes :

First Minute Plus Additional Minutes = Total Cost Of Call.

Example: 3.4 minute call. Assume 6 second increments billed at .0989 cents per increment and that your initial 30 second cost equals 1.5823 cents.

Step 1 — Add 30 second cost (1.5823 cents) to (.0989 cents x 5) to get first minute cost of 2.0768 cents or 2.08 cents rounded.

Step 2 — Subtract initial 1 minute from 3.4 minute total call (3.4 - 1 = 2.4 minutes left).

Step 3 — Convert the call to seconds. 2.4 minutes equals 144 seconds. Divide by 6 to get the number of 6 second intervals. 144/6 = 24 intervals.

Step 4 — Multiply 6 second intervals by 24. 24 x .0989 cents = $2.3736 or $2.37.

Step 5 — Add 1st minute cost to additional minute cost. $2.08 + $2.37 = $4.45.

Total cost of call should equal $4.45.

When auditing your long distance charges, pay particular attention to international calls billed at the 30 second minimum rate. The long distance carriers will often bill you the minimum 30 second rate even if the call is not completed. These calls can add up to a significant amount of money. Have your employees write down all international calls that are not completed or that have to be re-dialed because of bad transmission (crackling noises, static etc.). Credit should be obtained for these calls.

Bad transmission on domestic calls should also be tracked. Credit is usually given for poor transmission if the call is of short duration and the number in question is redialed within 24 hours.

THIRD NUMBER CALLS

A third number call takes place when someone calls a location but has the charges for the call billed to another location. Before the widespread use of telephone credit cards many salespeople used third number billing to make calls while they were on the road.

An example of how a third number call will appear on your bill follows:

DATE	TIME	NUMBER	PLACE	MIN	AMOUNT
Nov. 19	946AM	To Senegal	0221345555	55	$62.49
		From N.Y.	212 555-9999		

Someone from a New York location (212 555-9999) has dialed Senegal and charged their call to your telephone number. Typically these calls originate from a pay phone. If you see these calls on your bill and they are not authorized by you then you are entitled to credit for these calls (see Toll Fraud Auditing for more information on preventing toll abuse).

CREDIT CARD CALLS

Credit card calls should be checked very carefully. They are often compromised by "shoulder surfers". Shoulder surfers stand at pay phones and attempt to ascertain your credit card PIN number by watching as you dial it. They then set up call-sell operations where they resell your PIN number to others. Federal law holds you responsible only for the first $50 of usage. If your card is compromised, you are entitled to credit for the balance of the fraudulent usage.

PEAK AND DISCOUNT PERIODS

Most of the carriers offer off-peak calling discounts. Calls made during the normal work day are referred to as peak hour calls. Peak hours usually apply Monday to Friday 8 A.M. to 5 P.M. Calls made outside this corridor are discounted anywhere from 35% to 60% depending on the carrier. Hospitals and other 24 hour organizations should spot check their bill to make sure they are receiving off-peak discounts. To calculate your off peak discount follow the previous guideline on how to price a call and then multiply the cost of the call by the discount in effect. A discount of 60% means that you should only be billed 40% of the peak rate for the call. For example, a $3.08 cents peak call made at a night discount rate of 60% should cost (3.08 x .40) or 1.23.

Most carriers will also offer additional discounts to identified countries or locations. For example, AT&T offers an additional 10% discount to one selected foreign country under their most favored nation plan. It is important to insure that the country selected is the one you call the most frequently. You also need to spot-check calls to this country to make sure that the additional discount is properly applied.

MULTILOCATION, VOLUME AND VIP DISCOUNTS

Common long distance billing errors include discounts not worked and discounts promised but not properly applied. Long distance companies promise discounts, but once the sale is made, rarely verify that all your discounts are in effect. Most sales representatives are too busy working on their next sale to properly check your bill.

Most long distance companies reward high volume customers with additional discounts off their tariffed rates. These discounts are designed to stimulate usage. If properly applied, they can significantly reduce your bill.

Multilocation discounts allow you to consolidate the gross billing for all your locations in order to maximize usage based discounts. A multilocation discount is actually a volume discount that is applied to the gross usage of all your locations combined. (Multilocation discounts will be used synomously with volume discounts).

VIP discounts are applied when you agree to stay with a carrier for a certain time period. If, for example, you sign a contract guaranteeing that you will stay with a particular carrier for a year they will add additional discounts to your bill.

MCI's VISION service offers both Multilocation and VIP discounts, and as such, provides an excellent example on how these programs work. MCI is also an example of a sales driven company. Sales representatives are agressive, but are poorly trained on how to understand MCI's bills. They simply sell and move on. Discounts promised by these reps are often incorrectly applied.

VISION is MCI's premier service offering and is only offered to customers that bill $1,000 or more each month. Rates average 21 cents per call (actual charges: 21.2 cents minute for calls over 100 miles, 18.2 cents per minute for calls under 100 miles).

In order to audit the accuracy of an MCI VISION bill, you need to know MCI's discount schedule. VISION offers the following usage based volume discounts:

TOTAL MONTHLY SPENDING	DISCOUNT
1,000 To 9,999	10%
10,000 to 14,999	15%
30,000 and over	20%

If you sign a one year contract with MCI they will give you an additional discount called the VIP discount. This discount amount is also dependent on total monthly spending and is applied in addition to volume discounts.

ONE YEAR MCI VISION VIP	DISCOUNT
1,500 to 9,999	3%
10,000 to 14,999	5%
15,000 and over	7%

After each threshold is reached the VIP and volume discounts are calculated on the entire bill amount.

For record keeping purposes most companies request that each of their branch locations receive a separate invoice. Each branch location will usually sign off on the charges for their location and then forward their invoice to a central location for payment. These companies want and deserve usage discounts that are applied to the sum of the charges for all locations rather than to each location separately.

Example: Customer X has locations in New York, Miami, Houston, Boston and Chicago. Usage (domestic & international combined) per location is as follows:

Location	Total Usage
New York	$4,000.00
Miami	$1,500.00

Houston	$2,000.00
Boston	$1,900.00
Chicago	$3,599.00
Total	$12,999.00

All locations should be tied together under VISION's Multilocation Volume Discount Plan. To audit this customer's bill you need to know how these discounts should be applied by MCI.

The volume discount is first calculated by MCI against the sum total of all the locations and then each individual location is given a statement that details the per cent discount applied to their portion of the charges. This allows each location to remain responsible for its long distance usage while maximizing discounts. The following steps determine the correct discount application for customer A:

Step 1 — Determine the per cent volume discount due the customer (based on usage).

Total Usage = 12,999

First $9,999.00 of usage is discounted by 10%.

$9,999 x 10% = $999.90 credit.

Next $3,000.00 of usage is discounted 15%.

$3,000 x 15% = $450 credit

Take the total volume usage credit (1449.90) and divide by the total usage (12,999) = Per cent discount (11.15%).

Step 2 — Calculate the VIP discount. Use the same formula as previously used with the volume discount but substitute VIP discount rates.

First $9,999 x 3% = $299.97

Next $3,000 x 5% = $150

Take the total VIP credit ($449.97) and divide by total usage (12,999) = Per cent discount per location (3.46%).

Step 3 — Add up all the discounts. Total discount credit due customer is $1,899.87.

MCI now apportions the total discount by location according

to the usage at each location. Each location gets a multilocation discount of 11.15% and a VIP discount of 3.46%.

NEW YORK

USAGE		CHARGES
Total Usage		$4,000.00
Volume Discount	11.15%	$ 446.00 cr
VIP Discount	3.46%	$ 138.40 cr
Total Credit		$ 584.40 cr
Total Current Charges		$3,415.60

MIAMI

USAGE		CHARGES
Total Usage		$1,500.00
Volume Discount	11.15%	$ 167.25 cr
VIP Discount	3.46%	$ 51.90 cr
Total Credit		$ 219.15 cr
Total Current Charges		$1,280.85

HOUSTON

USAGE		CHARGES
Total Usage		$2,000.00
Volume Discount	11.15%	$ 223.00 cr
VIP Discount	3.46%	$ 69.20 cr
Total Credit		$ 292.20 cr
Total Current Charges		$1,707.80

BOSTON

USAGE		CHARGES
Total Usage		$1,900.00
Volume Discount	11.15%	$ 211.85 cr
VIP Discount	3.46%	$ 65.74 cr
Total Credit		$ 277.59 cr
Total Current Charges		$1,622.41

CHICAGO

USAGE		CHARGES
Total Usage		$3,599.00
Volume Discount	11.15%	$ 401.29 cr
VIP Discount	3.46%	$ 124.53 cr
Total Credit		$ 525.82 cr
Total Current Charges		$3,073.18

(Because of rounding, the sum of the credits for each location totals $1,899.16 as opposed to $1,899.87 when the discount is totaled in bulk).

An audit of this account found that the Boston and Houston locations were not receiving VIP discounts and that the Chicago location was not receiving either the volume or VIP discount. Because these locations were not included, the customer was paying too much for his long distance service at all the locations. The discounts for this customer was calculated incorrectly by MCI as follows:

Volume discount (excludes Chicago):

Total Usage = $9,400.00

9,400.00 x 10% =$940.00 credit

VIP discount (excludes Boston, Houston, and Chicago)

Total Usage = $5,500

5,500 x 3% =$165.00 credit

Chicago's Credit Calculated Separately:

Volume discount:

Total Usage =$3,599.00

3,599.00 x 10%=$359.90 credit

VIP discount

Total Usage=$3,599.00

3,599.00 x 3%=$107.97

Total Credit (Chicago only) =$467.87

Under this discount formula the customers' total credit is only $1,572.87 ($940.00 + $165.00 + $467.87).

To determine how much credit is owed this customer subtract the actual credit from the total credit due the customer.,

Total credit due customer (all discounts worked properly):

Volume discount	$1,449.90
VIP discount	$ 449.97
TOTAL	$1,899.87

Actual credit given:

Volume discount $1,299.90

VIP discount	$272.97
TOTAL	$1,572.87

Refund owed customer: $1,899.87 — $1,572.87 = $327.00

This refund amount has to be calculated independently for each month the discount was not properly applied. Taxes and interest should also be applied to any overcharges.

Many companies also have their credit card calls billed by location. These bills can and should be tied into your VISION volume and VIP discounts. Credit card usage should be tied into your regular long distance usage to maximize discounts. You should also make sure that your credit card calls are on your carrier's best service offering. For example, if your credit card calls are billed separately they may be placed on a service that is given to low volume users. MCI may bill your credit card calls at its Prism Plus rates instead of the more economical VISION rates. Besides paying more per call with Prism Plus you also pay a surcharge of 75 cents for each credit card call you make. VISION customers only pay 60 cents per credit card call. If your company is making 3,000 credit card calls per month and you are on Prism Plus instead of the VISION plan, you are overpaying $450.00 each month on surcharges alone. This overcharge is calculated as follows:

(3000 calls times 75 cents surcharge per call) $2,250 subtracted from (3000 calls times 60 cents surcharge per call) $1,800 or $450 monthly.

While MCI's discount plans are used for illustration, all carriers offer discount plans. The more complicated the plan, the greater the chance for error. An auditor should gather all contracts signed between the long distance carrier and his customer. All letters and agreements should be reviewed (if a bid was placed it should also be reviewed) to determine the discounts promised the customer. The auditor should then spot check each bill to find calls that meet the criteria for the discount. He should also calculate the rate for selected calls himself and compare his rate to the rate billed by the carrier. If the customer has multiple locations, the auditor should insure each location is receiving the proper discount.

LOCAL CALLING

Your local telephone company has a monopoly (in most areas) on intra-LATA calls. As the sole provider of service, they can charge whatever state regulators will allow. The absence of competition keeps rates artificially high. Long distance competition has resulted in a 40% decrease in the cost of an average long distance call since 1984.

Long distance rates are actually cheaper in many instances than the rates charges for local, intra-LATA calls. Consider that a three minute inter-LATA call placed from New York City to Los Angeles (a distance of 3000 miles) will cost you 63 cents using MCI's VISION service, while a three minute intra-LATA call placed from New York City to Eastern Suffolk (a distance of 100 miles) will cost you $1.03 under New York Telephone's Regional Calling Plan.

Competition in the local market is in its infancy. Local regulators know that competition reduces cost. They are, however, faced with the pressures that the large telephone companies can afford to place on them. Even so, most state regulatory agencies are slowly but surely introducing regulations that allow for more competition. Chapter 5 provides more information on the emerging intra-LATA competitors known as Competitive Access Providers or (CAPs).

In response to the threat of competition many local telephone companies have begun offering volume discounts and other programs to reduce the cost of local calling. As a bridge to full scale competition, your company should be utilizing these discounts.

BILLING FOR LOCAL CALLS

Most local telephone companies divide the LATAs they serve into calling bands. For example, New York Telephone divides LATA 132 (see Figure 4.4 on the following page), which comprises the New York Metropolitan Area into bands they call regions. The regions are as follows:

1. New York City
2. Lower Westchester County
3. Nassau County
4. Rockland County
5. Upper Westchester County
6. Western Suffolk County
7. Eastern Suffolk County

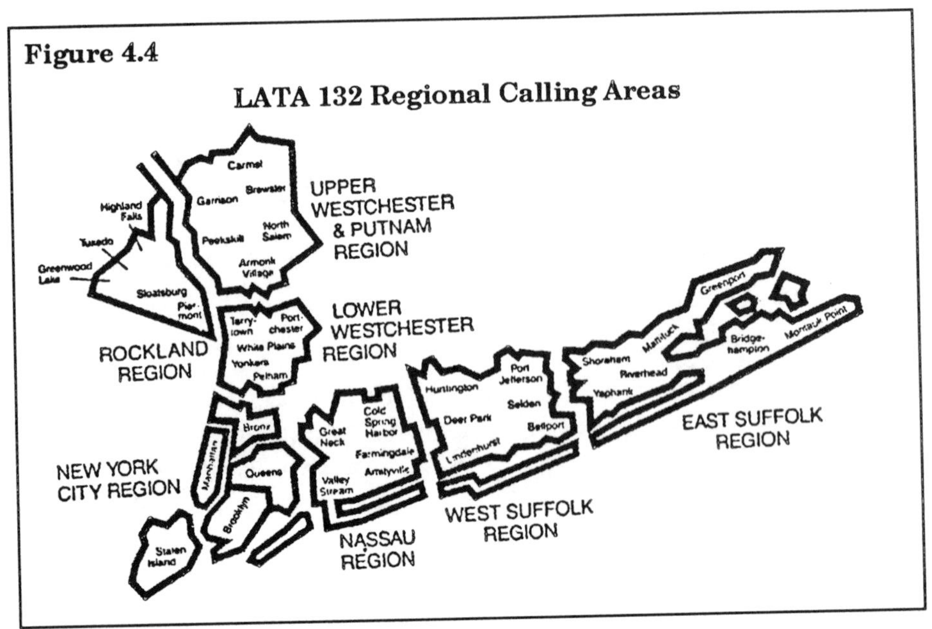

Figure 4.4

LATA 132 Regional Calling Areas

Calling rates charged by New York Telephone are region sensitive. The region you are located in is called your primary region. Calling regions that have physical boundaries with your region are known as adjacent regions. All other regions are called distant regions. Because these regions can be geographically far apart, calls between distant regions are sometimes called local long distance calls.

Rates for calls within your primary region are quite reasonable, while calls to distant regions are exorbitant. New York Telephone charges an initial rate for the first minute (within your primary area the initial period is three minutes) and then a lower rate for each additional minute. For a customer located in New York City, New York Telephone will charge the following for intra-LATA calls:

AREA	CALL REGION	INITIAL CHARGE	ADDITIONAL MINUTE
PRIMARY	NEW YORK CITY	8 CENTS/1ST 3 MIN	1.3 CENTS
ADJACENT	LOWER WESTCHESTER	14.7 CENTS/1ST MIN	4.6 CENTS
ADJACENT	NASSAU	14.7 CENTS/1ST MIN	4.6 CENTS
DISTANT	ROCKLAND	25 CENTS/1ST MIN	9.8 CENTS
DISTANT	UPPER WESTCHESTER	25.4 CENTS/1ST MIN	10 CENTS
DISTANT	WESTERN SUFFOLK	25.4 CENTS/1ST MIN	9.9 CENTS
DISTANT	EASTERN SUFFOLK	34.0 CENTS/1ST MIN	17.5 CENTS

Most local telephone companies offer discounts for off peak calling. The standard discount periods are based on the old Bell System standards:

40% - 9 P.M.-11 P.M. Monday through Friday and 5 P.M. to 11 P.M. Sunday.

65% - 8 A.M.-5 P.M. Sunday and 11 P.M.-8 A.M. everyday.

WAYS TO REDUCE YOUR LOCAL CALLING BILLS

Most local telephone companies offer both time and volume based discounts. The simplest way to reduce both your local and long distance bills is to implement a system that accumulates and sends your FAXs and electronic mail messages at night and other off peak times.

VIRTUAL WATS

Another way to reduce your local bills is to take advantage of volume based local calling plans. Most local telephone companies offer a form of discount local outbound WATs service. For example New York Telephone Company offers calls its outgoing WATS service, Virtual WATS. These services are really volume discount offerings and can save you money. New York Telephone's rates for Virtual WATS are on a sliding scale. The more calls you make, the cheaper the rate.

New York Telephone's Virtual WATS rates

Hours per month	Rate per minute
First 10 hours	16.6 cents
Next 90 hours	12 cents
Next 100 hours	10 cents
All additional calls	8 cents

There is no installation charge to convert to a Virtual WATS plan. There is, however, a minimum usage requirement of $100.00. The following example calculates the savings a major retail organization realized by switching to virtual WATS. The company has its showroom in New York City and its warehouse and corporate headquarters in Western Suffolk, Long Island, N.Y. The company makes, on average, 10,000 five minute calls each month to destinations within the Western Suffolk calling region. To determine how much a Virtual WATS plan will save this customer you must first calculate the cost of the calls under normal regional calling rates and then re-calculate the calls under Virtual WATS rates.

Step 1 — Determine the cost of the calls under New York Telephone's Regional Calling Plan. Calls to Western Suffolk are billed at a rate of 25.4 cents for the first minute and 9.9 cents for each additional minute.

10,000 calls times 25.4 cents (initial one minute rate) equals $2,540. 10,000 calls times 4 minutes times 9.9 cents per minute equals $3,960. Total cost for the calls equals $6,500.

Step 2 — Determine the cost per minute for a call.

$6,500 divided by 50,000 (total minutes) equals 13 cents per minute.

Step 3 — Determine the cost of the same calls under a Virtual WATS program.

First 10 hours (600 minutes of usage)	$100
50,000 minus 600 equals 49,400	
Next 90 hours (5,400 minutes of usage)	$ 648
49,400 minus 5,400 equals 44,000	
Next 100 hours (6,000 minutes of usage)	$ 600
44,000 minus 6000 equals 38,000	

Next 38000 minutes of usage (8 cents per min)	$3,040
Total Virtual WATS billing	$4,388

Step 4 — Determine monthly savings. $6500 minus 4,388 equals a monthly savings of $2112.

FOREIGN EXCHANGE (FX) LINES

Dependant on your location and calling patterns an FX line can also save you money and give you a presence in another calling area. An FX line provides you with a telephone number that is not local to your central office. FX lines are commonly used by companies that do business in regions that have a different area code from the one they are in. Companies use FX lines because many potential customers are reluctant to dial a number outside their own area code.

An FX line consists of a dedicated special service line from your local CO to the CO that is in the area you wish to establish a presence in (see figure 4.5). You pay a fixed monthly fee for the use of the special service circuit that connects you to the foreign CO. Central to your monthly fee is mileage charges calculated on the distance between your local CO and the foreign CO. Outgoing calls are charged at the local rate in effect at the foreign location. The following provides an example of how an FX line can reduce your bill.

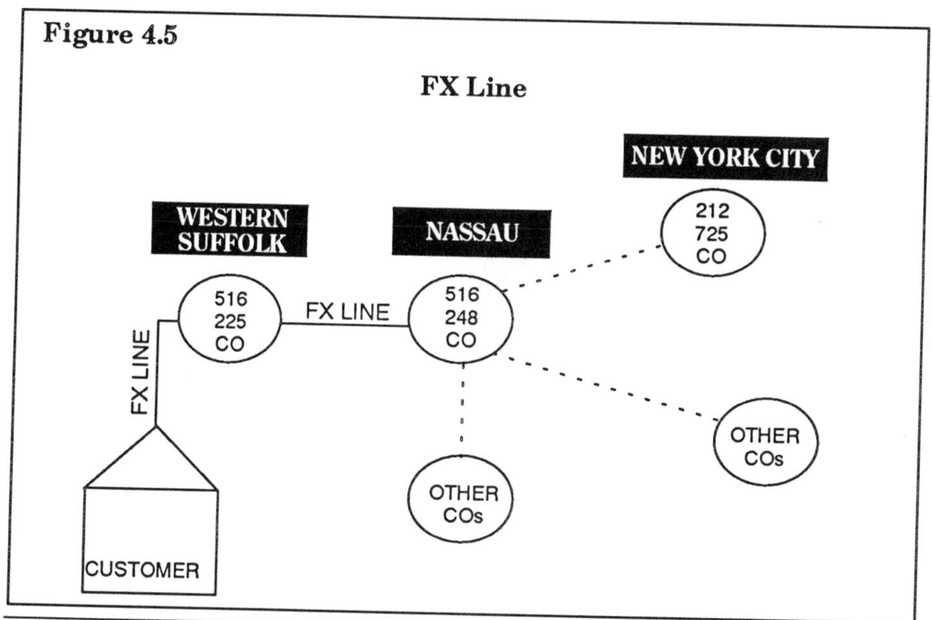

Example - Customer B is physically located in Western Suffolk and makes a high number of calls to Nassau County and New York City. On an average month this customer makes 5,000 five minute calls to Nassau County and 3,000 five minute calls to New York City. Under the standard regional calling plan the calls will be billed at the following rates:

REGION	RATE/1st MIN	ADD'L MIN
Nassau	14 cents	4.5 cents
New York City	25.4 cents	9.9 cents

From this customer's local CO, Nassau County is an adjacent region while New York City is a distant region.

To determine the cost of Nassau County calls:

Multiply 5000 by the initial one minute rate of 14 cents. Subtotal $700. Take the 20000 additional minutes and multiply by the additional minute rate of 4.5 cents. Subtotal $900. Total charge $1,600.

New York City calls:
Multiply 3000 by the initial one minute rate of 25.4 cents. Subtotal $762. Multiply the additional minutes (12000) and multiply by the additional minute rate of 9.9 cents. Subtotal $1,188. Total cost for New York City calls $1,950.

Total cost for all calls $3,550.

An FX line to a CO located in Nassau County will alter the billing rates for your calls. All calls made over the FX line will be billed at Nassau County's regional rates as follows:

REGION	RATE/MIN	ADD'L MIN
Nassau	8 cents for the 1st 3 minutes	1.3 cents
New York City	14.7 cents for the 1st minute	4.6 cents

The main difference is that calls to Nassau County are now considered primary area calls. As such, the initial rate of 8 cents covers the first three minutes. To calculate your monthly savings, re-calculate your monthly charge for local calls under the new rates.

Step 1 — Determine the cost of the initial period for Nassau County calls. 5000 calls at 5 minutes equals 25000 minutes. The initial period cost is determined by multiplying 15000 by 8 cents ($450). Subtract the total initial minutes (5000 calls x 3 minute initial period or 15000 minutes total) from the total number of minutes (25000). Multiply 10000 additional minutes by 1.3 cents to get a subtotal of $130).

Total cost of calls placed within Nassau County equals $580.

Step 2 — Calculate the cost of calls to New York City. 3000 calls at initial 1 minute rate of 14.7 cents equals $441.00. Multiply additional minutes (12000) by 4.6 cents which equals $552. Total cost of local calls placed to New York City equals $993.

Step 3 — Determine total cost of all local calls:

$580 plus $993 equals $1,573.

Step 4 — Determine your monthly savings on calls by subtracting the cost of the calls placed over the FX line versus calls placed over POTS lines billed at regional calling rates. $3,550 minus $1,573 equals a cost savings of $1,977 monthly.

To determine actual cost savings, you must use the principles of application engineering. To justify this volume you will need at least five FX lines. Therefore, to calculate true cost savings, you must first determine the monthly cost of an FX line as well as the initial installation charge per FX line to determine your break-even point. It is important that your FX line terminates in the closest foreign CO that allows a cheaper rate structure. If your FX line terminates in a CO that is further away than is necessary you will pay additional monthly mileage costs. Follow the following steps to determine your break-even point.

Step 1 — Determine the monthly cost of an FX line.

Assume this customer's local CO in Western Suffolk is 4 miles from the nearest CO located in Nassau County. Typical monthly charges for an FX line include:

QUANT	ITEM	DESCRIPTION	RATE
1	FYW	Foreign Exchange Terminal Charge	13.86
1	TGX	Trunk Charge	16.23
1	TJB	Touchtone Trunk	4.87

16	1LJBY	16 Quarter Miles	87.58
1	9ZR	FCC Line Charge	5.48
TOTAL			128.02

Five FX lines will cost a total of $640.10 per month.

Step 2 — Determine your actual monthly savings by subtracting the monthly cost of the FX lines from your monthly savings on your calls. $1,977 minus $640.10 equals $1,336.90.

Step 3 — Determine installation costs: The installation cost for five FX lines is as follows:

1. (1) Service Charge	$ 56.00
2. (1) Premise Visit Charge	$ 19.00
3. (5) Line Charges	$ 358.75
4. (5) Channel Connection Charges	$2,366.60
Total Installation Charges	$2,800.35

Step 4 — Determine your break-even point by dividing the installation cost by your monthly savings. $2,800.35 divided by $1,336.90 equals 2.09 months. Installing the FX lines pays for itself in just over 2 months.

By installing 5 FX lines you also add additional capacity to your firms network and make it more convenient and cheaper for customers to call you. These additional FX lines satisfy the ultimate objective of Applications Engineering which is to save you money and simultaneously increase the quality and capacity of your network.

LARGE VOLUME DISCOUNT PLANS

As competition slowly evolves in the marketplace you will see the local telephone companies begin to offer a variety of local calling volume discount plans. As competition will initially focus on very large corporations these discount plans will initially cater to very large corporations. In areas like New York City, which are heavily targeted by CAPs, New York Telephone must develop plans to keep large customers on their network. An example of a discount for large customers is New York Telephone's LARGE VOLUME DISCOUNT PLAN (LVDP). This plan was tariffed on 7/24/92. The highlights of the plan include the following:

1. The customer must commit a minimum of $500,000 of applicable intra-LATA billed usage per year.

2. The plan applies to the sum of applicable local call revenue, state-wide for the subscriber's specified billing numbers (BTNs).

3. The plan includes only directly dialed calls made as follows:

A. Non-adjacent inter-region Regional Call Plans.
B. Upstate New York intra-LATA toll calls.
C. WATS 800 intra-LATA calls.
D. 800 VALUFLEX calls.

4. The plan excludes all calls made under any other New York Telephone discount plan like Virtual WATS. It also excludes calls to interactive services and mass announcement services (i.e. 700,800, 900 and 976 services).

5. The discount rates are based on the amount of calls that you make. To qualify you must commit to a certain minimum usage range as follows:

ANNUAL USAGE COMMITMENT	DISCOUNT
$500,000 to $749,999.99	16%
$750,000 to $999,999.99	17%
$1,000,000.00 to $1,249,999.99	18%
$1,250,000.00 to $1,449,999.99	19%
$1,500,000.00 to $1,749,999.99	20%
$1,750,000.00 to $1,999,999.00	21%
$$2,000,000 and above	22%

If the customer does not meet the minimum usage requirements the full local calling tariff rates are retroactively applied. If the customer exceeds his commitment then a higher discount will be applied retroactively.

6. A one time record order charge of $35.00 applies to each main billing number placed on the plan. For example, a bank with 100 branch offices, would pay at least 100 times $35 or $3,500 to take advantage of this plan. (the actual charge will probably be higher as some branch offices will have more than one main billing number per branch).

This plan only applies to extremely large users that have many locations. It is geared toward government and banks with extremely high volume usage to non-adjacent regions. The only advantage this plan offers when compared to Virtual WATS is that a company can bundle its low volume locations with its high usage locations to get discounts. The concept of the LVDP is similar in nature to AT&T's SDN service. This program only marks the beginning of the war for the local market. Chapter 5 provides additional information on how competition in the local market will effect your bill.

CHAPTER 5

COMPETITION IN THE LOCAL LOOP

COMPETITIVE LOCAL CARRIERS (CLCs)

Competitive Local Carriers, like the Teleport Communications Group Inc., were originally conceived and economically justified as bypass companies. In fact, they were originally known as Competitive Access Providers or CAPs. The term competitive local carrier is applied to CAPs that offer, or seek to offer, traditional telco services like switched calling capabilities.

CLCs connect companies directly to their long distance carrier, thereby bypassing the local telephone company's CO. The avoidance of the local telephone company reduces the overall cost of long distance service for a customer utilizing their services.

The long distance carrier does not have to pay access charges to the local telephone company and can therefore charge a cheaper rate to the end user. As a way of recovering lost revenue, the local telephone company also will also provide you with a T1 that bypasses the switched network.

In addition to bypass services, the CLCs also offer private lines, network management, local area interconnection and some switched services at prices equal to or better than those offered by the established local telephone companies. Competitive Local Carriers also provide your telephone network with the benefits of redundancy and disaster avoidance. The smart manager will seek to diversify his network so as not to put all his telecommunications eggs into one basket.

When a customer agrees to dedicated T1 service to a long distance carrier's POP, the long distance carrier will often order the intra-LATA T1 on behalf of the customer. Long distance companies prefer to use CLCs where possible. Their goal is to economically strengthen the CLCs because they see the Bell Operating Companies as potential long distance rivals.

Although the local Bell Operating Companies are currently prohibited from providing long distance service, they are constantly lobbying congress to pass legislation that will overturn this prohibition. The long distance carriers naturally wish to avoid strengthening these powerful, potential rivals.

In contrast to the Bell Operating Companies, the Teleport Telecommunications Group (TCG) has assured the long distance carriers that it does not seek to compete against them..

Thanks to recent favorable rulings by the FCC, and other favorable rulings by various State Public Utility Commissions, the CLCs have extended their network offerings and the areas they are able to serve.

TELEPORT'S LOCAL CALLING SERVICES

TCG maintains networks in Boston, Chicago, Dallas, Houston, Los Angeles, New York and San Francisco. In addition to these established networks, TCG is also constructing networks in many other cities.

TCG provides local calling service (in direct competition to the local Bell companies) to customers that are connected to its fiber optic network. Figure 5.1 on the following page depicts a simplified Teleport network configuration.

The best example of local calling competition, exists within the New York Metropolitan Area. TCG's fiber optic network is illustrated by Figure 5.2 on the following page. The Teleport serves this area with 2 5ESS switches. Each switch has the

FIGURE 5.1

TCG NETWORK CONFIGURATION

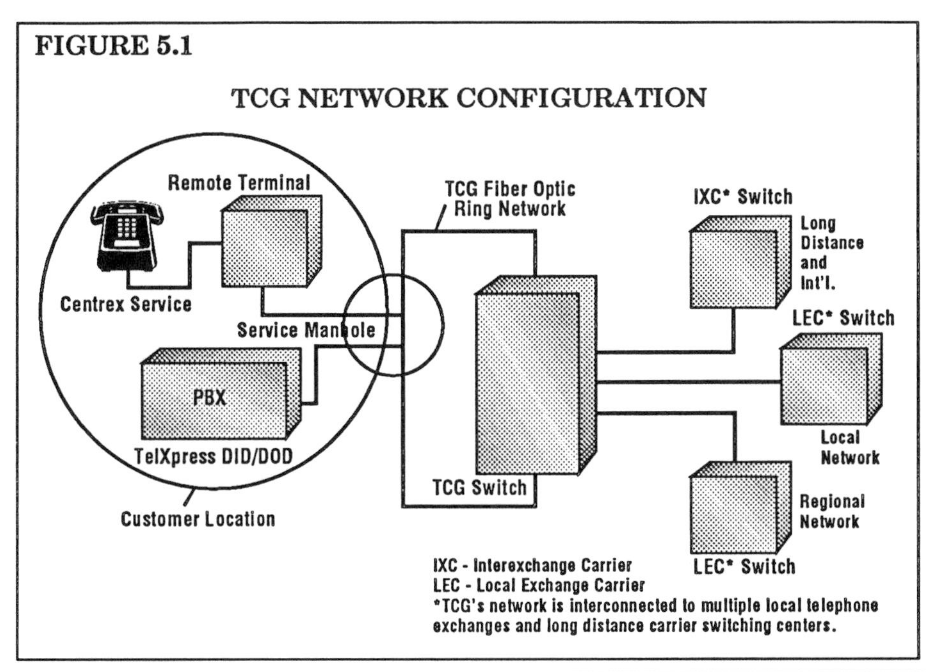

FIGURE 5.2

TCG FIBER NETWORK

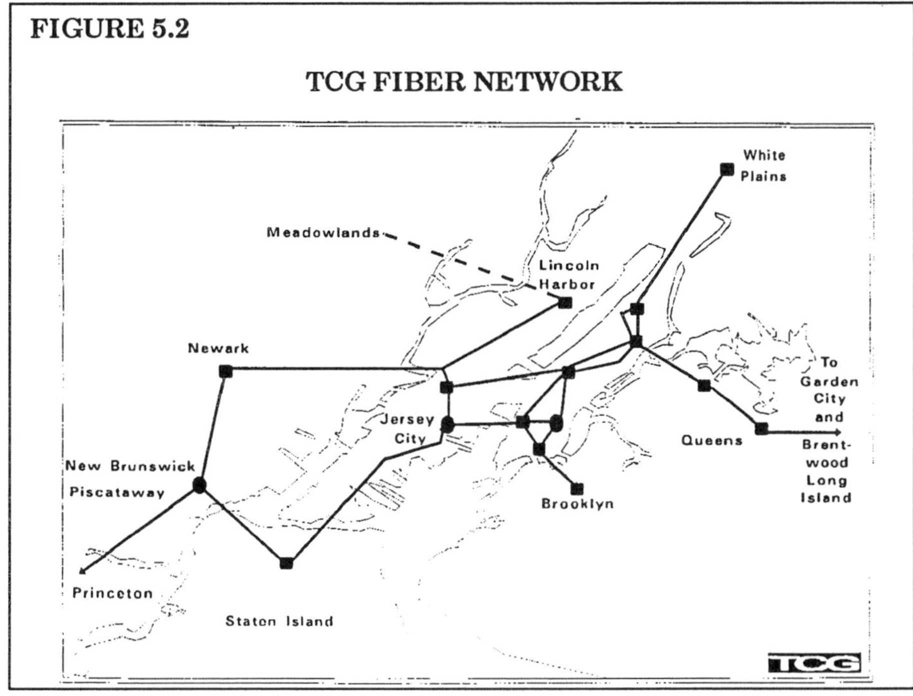

capacity to serve 100,000 lines. These switches allow TCG to provide local telephone service in direct competition with New York Telephone.

TCG maintains the same regional calling zones within LATA

132 (New York Metropolitan Area) as does New York Telephone. It also utilizes the same discount periods as New York Telephone. The following compares regional calling rates that would be charged to a customer located on Wall Street in New York City:

LOCAL CALLING: TELEPORT VERSUS NEW YORK TELEPHONE COMPANY

CALLS WITHIN NEW YORK CITY

TELEPORT	**NEW YORK TELEPHONE COMPANY**
8 CENTS/1ST 3 MINUTES	8 CENTS/1ST 3 MINUTES
1.3 CENTS EACH ADDITIONAL MINUTE	1.3 CENTS EACH ADDITIONAL MINUTE
COST OF A 5 MINUTE CALL: 10.6 CENTS	COST OF A 5 MINUTE CALL: 10.6 CENTS

CALLS TO NASSAU COUNTY, NEW YORK

TELEPORT	**NEW YORK TELEPHONE COMPANY**
12 CENTS/1ST MINUTE	14.7 CENTS/1ST MINUTE
3.8 CENTS EACH ADDITIONAL MINUTE	4.6 CENTS EACH ADDITIONAL MINUTE
COST OF A 5 MINUTE CALL: 27.2 CENTS	COST OF A 5 MINUTE CALL: 33.1 CENTS

CALLS TO WHITE PLAINS, NEW YORK

TELEPORT	**NEW YORK TELEPHONE COMPANY**
12 CENTS/1ST MINUTE	14.7 CENTS/1ST MINUTE
3.8 CENTS EACH ADDITIONAL MINUTE	4.6 CENTS EACH ADDITIONAL MINUTE
COST OF A 5 MINUTE CALL: 27.2 CENTS	COST OF A 5 MINUTE CALL: 33.1 CENTS

VIRTUAL WATS PLAN

TELEPORT	**NEW YORK TELEPHONE COMPANY**
FIRST 10 HOURS:15 CENTS/MINUTE CENTS/MINUTE	FIRST 10 HOURS:16.7 CENTS/ MINUTE
NEXT 90 HOURS: 9.9 CENTS/MINUTE	NEXT 90 HOURS:11 CENTS/MINUTE
NEXT 100 HOURS: 8.1 CENTS/MINUTE	NEXT 100 HOURS:9 CENTS/MINUTE
OVER 200 HOURS:6.3 CENTS/ MINUTE	OVER 200 HOURS:7 CENTS/MINUTE
COST OF 1000 HOURS OF USAGE: $4,135	COST OF 1000 HOURS OF USAGE: $4,594.

As you can see, significant savings are available from TCG, particularly with its virtual WATS service.

NEW YORK/NEW JERSEY CORRIDOR CALLS

Calls between Northern New Jersey and New York City can be completed either by a long distance carrier, your local telephone company or by TCG. This area provides us with a preview of a future where full scale competition will drive down prices and increase the quality of service.

The 1982 Divestiture Agreement allows New York Telephone to bill calls that originate in New York City and terminate in Northern New Jersey, in spite of the fact that the call crosses LATA boundaries. To place the call over New York Telephone's network, the caller must prefix the call with the access code 10NYT. Conversely, calls that originate in Northern New Jersey and terminate in New York City can be placed over New Jersey Bell's network by prefixing the call with 10NJB.

The following compares rates charged for calls placed within the New York/New Jersey Corridor:

1. NEW YORK TELEPHONE COMPANY*	11.26 CENTS PER MINUTE
2. TCG*	11.9 CENTS PER MINUTE
3. NEW JERSEY BELL	14.4 CENTS PER MINUTE
4. MCI	18.2 CENTS PER MINUTE

*Additional discounts apply if used with virtual WATS plans.

COLLOCATION

Collocation allows CLCs to rent space within a Bell Operating Companies' Central Office. The CLCs can then use this space to set up and maintain their own fiber optic facilities and equipment. More importantly, it allows CLCs to directly connect their facilities to the local telephone company's network. Collocation makes it easier for CLCs to reach and supply special service circuits to customers outside their existing fiber network.

In September of 1992 the FCC ruled that collocation must be provided to all CLCs throughout the United States. This decision provides the basis of a competitive special access marketplace. To illustrate the impact collocation will have on special access billing, consider how the telephone company currently bills itself for special access services.

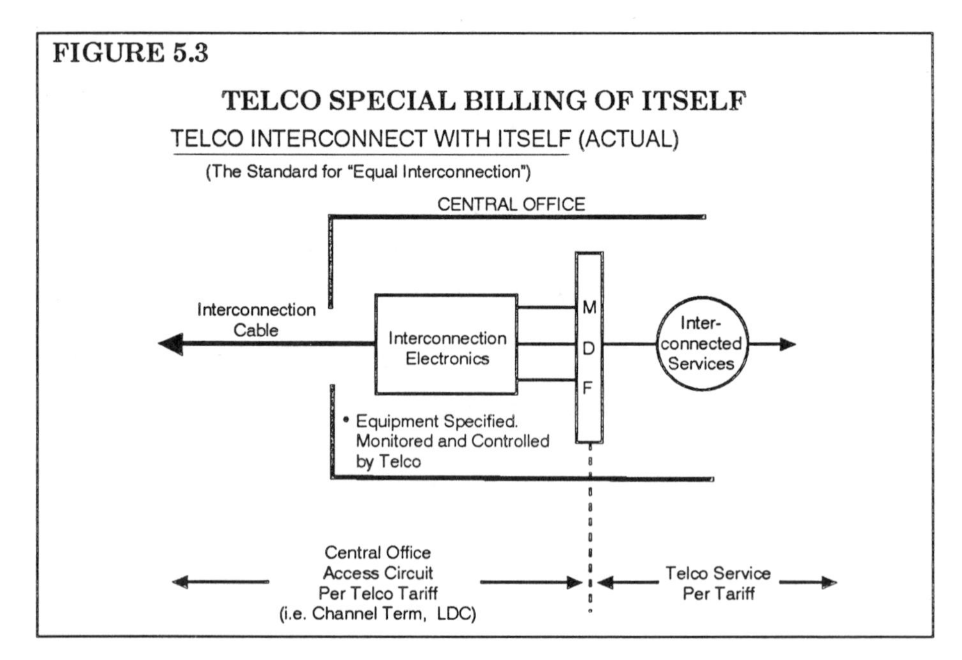

Figure 5.3 details how the telephone company connects special service circuits (that are ordered and provisioned by its own employees) to its own network.

Figure 5.4 illustrates how many telephone companies currently connect CLC originated special service circuits to their networks.

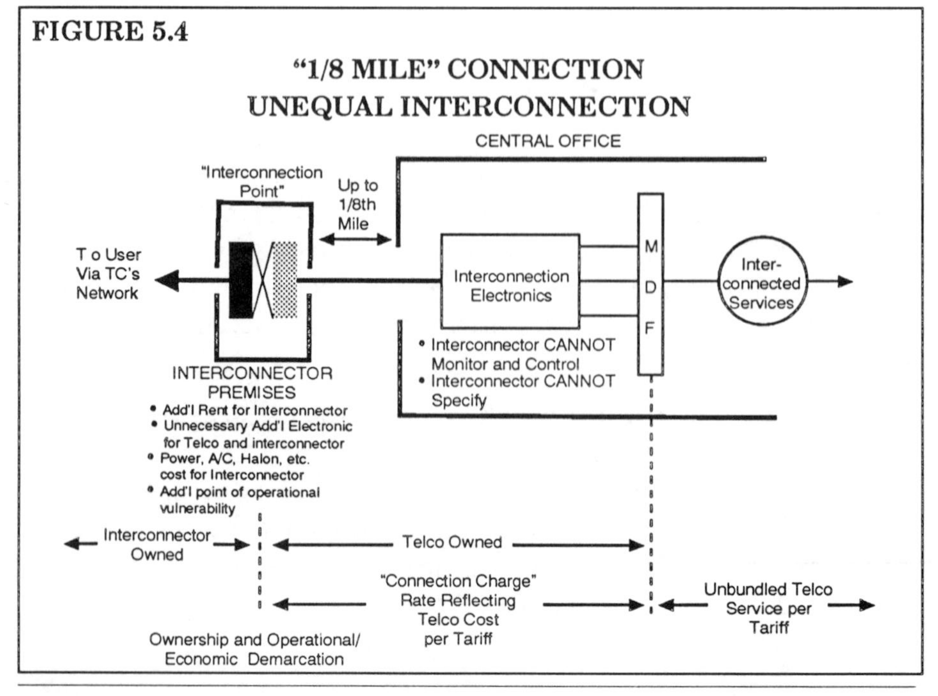

In figure 5.4, the point at which the CLC's facilities interconnect with the telephone company's network is approximately 1/8 mile from the telephone company's CO. The CLC must pay mileage charges to the local telephone company based on this distance.

Physical collocation is now a reality in many areas. For example, New York Telephone tariffed physical collocation on May 10, 1991. This tariff is known as OTIS II. Figure 5.5 illustrates the impact collocation has on the billing of the CLCs. You will notice that since the CLCs facilities are located directly at the CO, (mileage equals zero) mileage charges no longer apply.

FIGURE 5.5

PHYSICAL CLC CONNECTION TO TELCO NETWORK

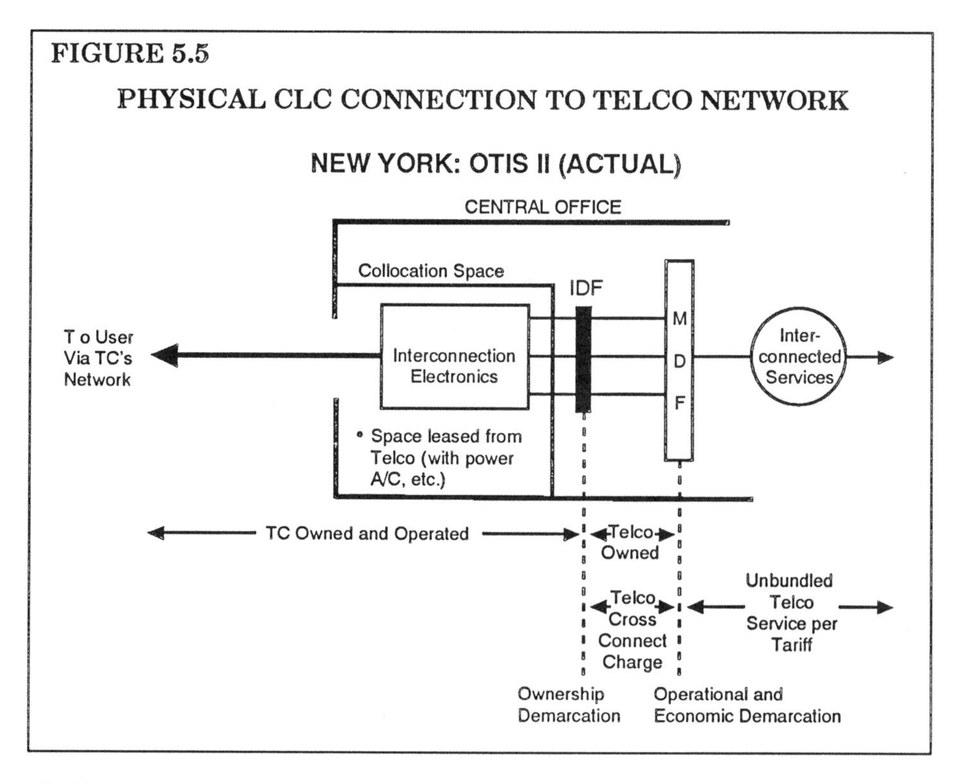

Collocation is a major step toward full scale special access competition. In the short term, however, many customers will not be able to take advantage of the benefits of the FCC decision. As discussed in chapter 3, most local telephone companies have been offering discounts to customers that sign long term commitments. The FCC ruling only allows customers with term plans in excess of three years to switch to a competitor without penalty. It is estimated that within NYNEX's territory (New York and New England) 75% of the DS-3 term arrangements are for exactly three years.

Competition in the local marketplace is moving forward at a rapid pace. Before you sign a term plan with your local carrier, consider all the potential ramifications.

UNBUNDLING OF THE LOCAL LOOP

The last great hurdle to full scale competition in the local market will be the unbundling of the local loop. Unbundling is defined as the division of a service into its elements, and the assignment of a separate price for each element.

Prior to the Divestiture Agreement, you simply called the telephone company and ordered phone service. The telephone company representative quoted you one monthly rental price which included the cost of the line to the central office, touch tone, inside wiring, jack charge(s) and the rental of a telephone set. Figure 5.6 illustrates this configuration. Since the 1982 Divestiture Agreement prohibited the local telephone company from providing telephone sets, the cost of telephone service had to be separated or unbundled.

FIGURE 5.6

PRE-DIVESTITURE TELEPHONE BILLING SCHEMATIC

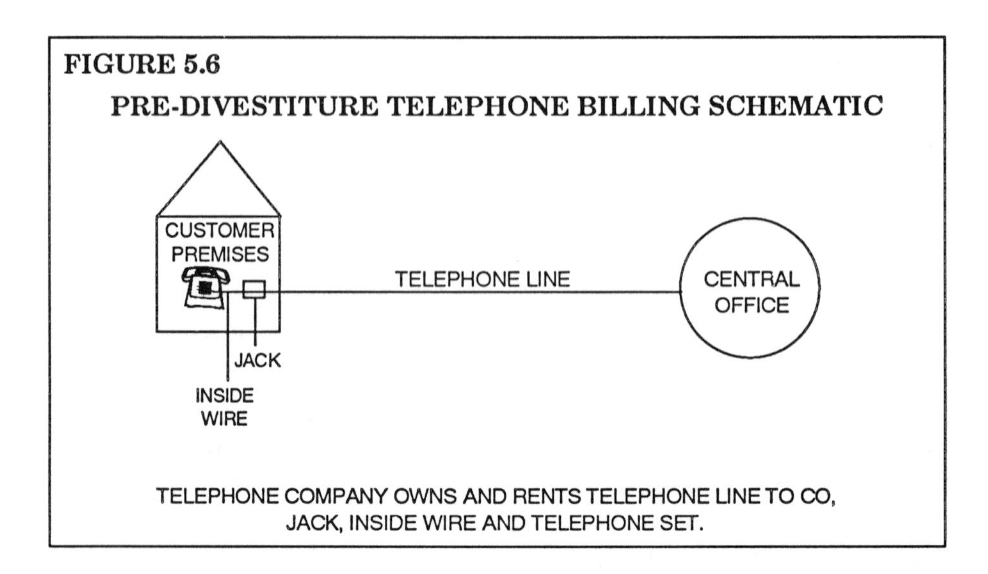

TELEPHONE COMPANY OWNS AND RENTS TELEPHONE LINE TO CO, JACK, INSIDE WIRE AND TELEPHONE SET.

To accomplish this separation the local telephone companies set up a demarc point at which all lines and equipment on one end of the demarc is owned and maintained by the telephone company. Equipment on the other side of the demarc is customer owned and maintained. Figure 5.7 illustrates this billing arrangement.

FIGURE 5.7

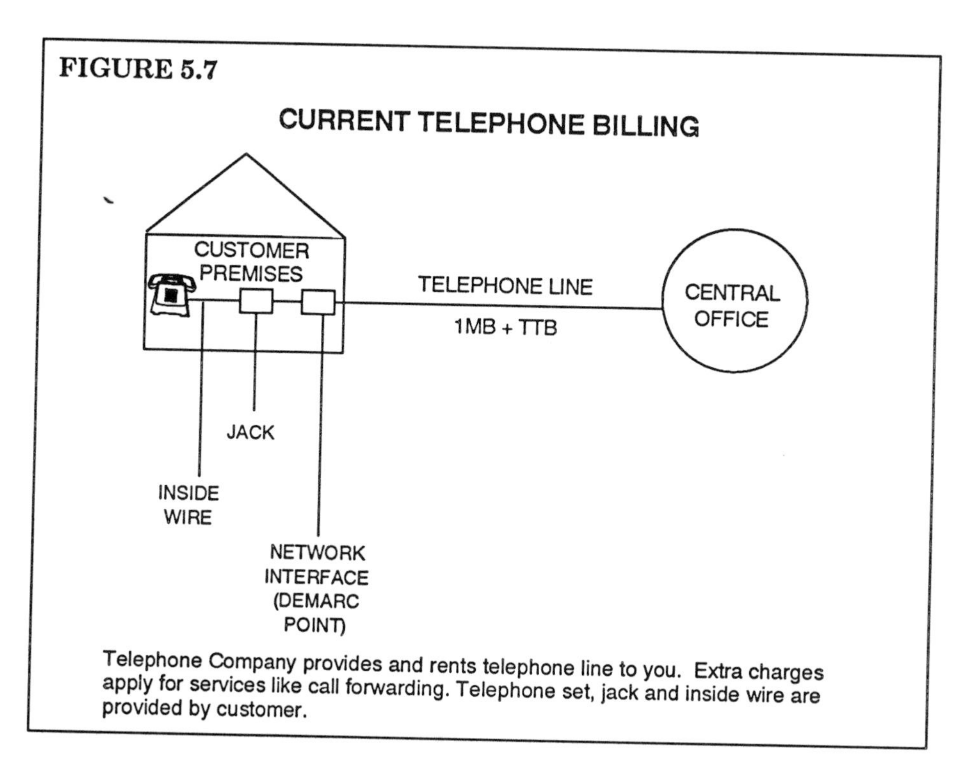

Telephone Company provides and rents telephone line to you. Extra charges apply for services like call forwarding. Telephone set, jack and inside wire are provided by customer.

Full scale competition requires the further unbundling of the telephone company side of the demarc. Currently, your telephone line to the CO is billed as a single entity. It actually includes two components: the line or link to the central office and the actual connection (port) into the telephone switch.

To realize full scale competition, a CLC should be able to supply you with a link to the telephone company's switch. This would allow a CLC to connect you to the local network. The CLC would pay the local telephone company a charge to obtain dial tone from its switch, and would bill the end user for entire cost of the service. Figure 5.8 on the following page illustrates this arrangement.

FIGURE 5.8

UNBUNDLING OF THE LOCAL LOOP (PROPOSED)

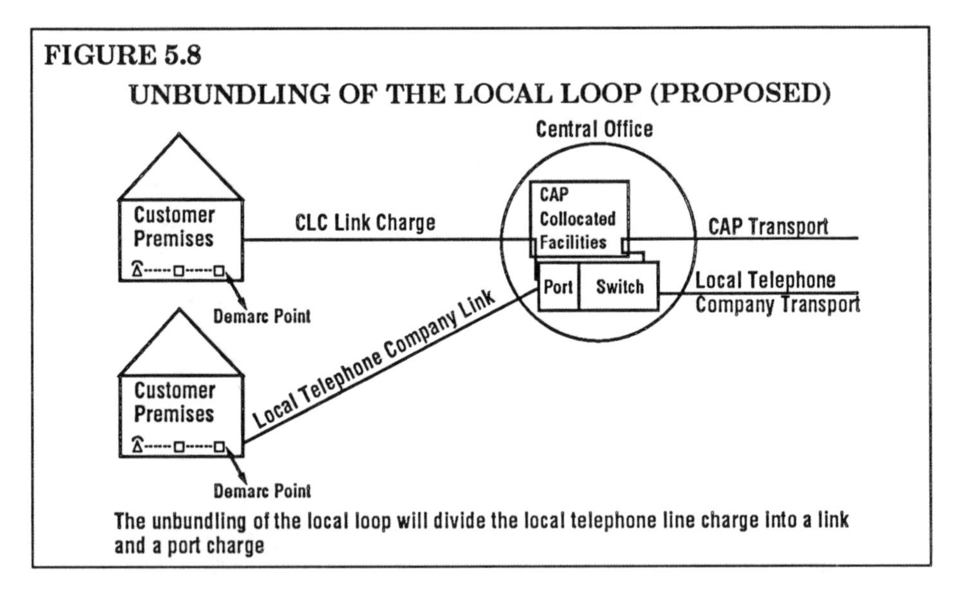

The unbundling of the local loop will divide the local telephone line charge into a link and a port charge

NEW YORK TELEPHONE'S UNBUNDLING TARIFF

On November 25, 1991 New York Telephone filed a tariff that would unbundle rates for basic access service. This tariffs offers a preview of the effect unbundling will have on future telephone bills. The tariff filed by New York Telephone separates the current cost for a CO line into a link and a port charge.

As detailed in Chapter 1, you pay a monthly rental charge of $16.23 (in New York City) for the line that runs between your business and the local CO. This line connects you to the telephone company's switch and allows you to dial just about anywhere in the world. The USOC code for this line is 1MB.

Under New York Telephone's proposed unbundling tariff, this combined charge would be separated into two separate charges. The first charge would be for $12.12 and would retain the USOC code 1MB. The second charge would be the port connection charge billed at $5.08 monthly. The USOC for this charge would be BPAL. Figure 5.9 depicts this connection.

You will notice that under this proposal the total cost of a single business line from New York telephone Company would increase from $16.23 to $17.20 ($12.12 plus $5.08).

FIGURE 5.9

N. Y. TELEPHONE - PROPOSED UNBUNDLED USOC BILLING

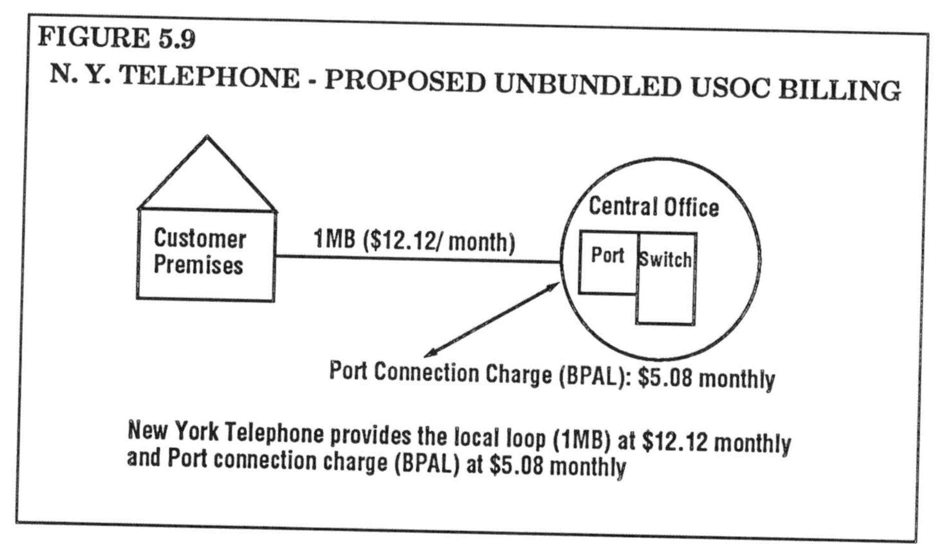

This unbundling tariff would allow CLCs to provide the link that runs from the customer's premise to the local telephone company's CO. The CLCs would then make separate arrangements to connect each end user to the telephone company's switch via the port charge. The CLC would provide the end user with a single bill for telephone service.

SUMMARY OF CHANGES THAT MUST BE MADE FOR EFFECTIVE LOCAL SERVICE COMPETITION

According to TCG, nine prerequisites are required to realize full scale competition.

Physical Interconnections

1. Cost-based unbundled local loops between a subscriber's premises and the telephone companies' local serving offices. (New York Telephone's proposed unbundling tariff seeks to address this perquisite).

2. Cost based central office interconnection arrangements, and cost based connections between the unbundled, cost based loops and the central office (CO) interconnection arrangement. Many states currently require physical collocation. The FCC ruling in September 1992 makes physical collocation a nationwide requirement.

3. Local telephone number portability, so that subscribers

can exercise free consumer choice by changing from one local exchange carrier to another without sacrificing their existing telephone numbers. (similar to 800 number portability).

Logical Interconnections

4. Equal access to and equal status in the local telephone company Signalling System 7 Systems, including databases and network routing processes (i.e. , the local telephone companies' routing systems treat TCG's class 5 switches no differently from their own Class 5 end office switches).

5. Equal access to and equal status in the local telephone companies' tandem switching and interoffice networks.

6. Integration of TCG's competing Class 5 and Class 4 switches into the local telephone companies' local routing plans, with the integrating accomplished through unbundled switching and facility elements at cost-based rates.

Financial and Administrative Interconnections

7. Unbundled, cost-based local telephone company rates, to terminate local calls originating on TCG's systems that are delivered on TCG's network to the local telephone companies' Class 5 local serving office or tandem switch (i.e., a local service "access charge").

8. Payment by the local telephone companies' for TCG's termination of local telephone company originated local calls on TCG's systems.

9. Cooperative engineering, operational, maintenance and administrative practices and procedures.

In addition to the previous prerequisites, full scale competition can only be accomplished if Bell Communications Research (Bellcore) and the National Carriers Association (NECA) treat TCG in the same manner as they treat any independent (non-Bell) local exchange service carrier. Equal treatment by Bellcore and NECA would have to extend to such critical functions as:

—Administration of the North American Numbering Plan.

—Dissemination of Network Routing Guides.

— Coordination of technical, operational and administrative standards and practices.

TCG's major competitor is Metropolitan Fiber Systems, Inc. (MFS). MFS also maintains competitive telephone networks throughout the United States. When considering special access services, contact TCG, MFS as well as the local telephone company. This will allow you to compare services and will also allow you to select the service that best meets your needs.

There are a multitude of CLCs throughout the United States, and the list keeps growing. Some CLCs are local to a particular city or state, while others like TCG and MFS are nationwide. The concerned billing specialist will contact these companies and gather price and service information.

The following lists some of the major CLCs and the areas they serve.

COMPANY	REACH NUMBER	AREA SERVED
ACC CORP.	716 987-3000	ROCHESTER, N.Y.
ASSOCIATED COMMUNICATIONS	213 387-9271	LOS ANGLES, CA.
BAY AREA TELEPORT	800 621-5003	SAN FRANCISCO, CA.
CITY SIGNAL INC.	616 235-4990	DETROIT, MICH. GRAND RAPIDS, MI.
DIGINET INC.	312 663-8200	CHICAGO, IL. MILWAUKEE, WI
DIGITAL DIRECT INC.	214 744-0190	DALLAS, TX. SEATTLE, WA.
EASTERN TELELOGIC CORP.	215 337-8899	PHILADELPHIA, PA.
FIBERNET, INC.	716 454-6990	ALBANY, N.Y. BUFFALO, N.Y. ROCHESTER, N.Y. SYRACUSE, N.Y.
INDIANA DIGITAL ACCESS CORP.	317 849-5639	INDIANAPOLIS, IN.
INSTITUTIONAL COMMUNICATIONS	703 827-5995	WASHINGTON, D.C.
INTERMEDIA COMMUNICATIONS	305 470-2424	MIAMI, FLA. ORLANDO, FL. TAMPA, FLA.
LOCATE	212 509-5115	NEW YORK, N.Y. CHICAGO, IL. BOSTON, MA. DETROIT, MI. WASHINGTON, D.C.

METROCOMM, INC.	614 221-9230	COLUMBUS, OH.
METROPOLITAN FIBER SYSTEMS	708 218-7200	CHICAGO, IL.
		BOSTON, MA.
		BALTIMORE, MD.
		DALLAS, TX
		HOUSTON, TX
		LOS ANGELES, CA.
		MINNEAPOLIS, MI.
		NEW YORK, N.Y.
		PHILADELPHIA, PA.
		PITTSBURGH, PA.
		SAN FRANCISCO, CA.
MITEL DIGITAL DISTRIBUTION	714 833-7171	LOS ANGELES, CA.
MWR TELCOM INC.	515 242-4360	DES MOINES, IO
NORTHEAST NETWORKS	914 428-7303	WESTCHESTER, N.Y.
TELEPORT COMMUNICATIONS GROUP, INC.	800 628-5608	BOSTON, MA.
		CHICAGO, IL.
		DALLAS, TX.
		HOUSTON, TX.
		LOS ANGELES, CA.
		NEW YORK, N.Y.
		SAN FRANCISCO,CA

THE FUTURE

Competition in the local loop is moving to the suburbs. CLCs are targeting business parks and offices that ring many major cities. For example, Northeast Networks Inc., is in the process of building a 30 mile fiber optic network in Westchester County, N.Y.(a suburb of New York City). As the cost of installing fiber drops, more and more areas will have access to competitive local service. In response to the increasing competition, the Bell companies are reducing their prices. New York Telephone recently reduced the cost of mileage on T1's by 85%.

Telephone bill auditors and Application Engineers face a happy future. Competition in the local loop will bring many opportunities for businesses, but it will also bring confusion.

Customers will receive multiple bills from multiple local carriers. As companies switch between local carriers, overbilling is sure to occur. If, for example, you switch some of your T1s to TCG from New York Telephone, you must verify that your monthly bill from New York Telephone reflects the proper reduction. You must also verify that TCG bills you correctly for the T1s provided by them.

The principles of Applications Engineering will help you accurately compare services between the local carriers and determine your true cost savings. While competition offers many cost benefits, it also increases the responsibility placed on the end user. You must become a smart shopper and an educated consumer to take advantage of the emerging world of local service competition.

GLOSSARY

Advance Billing — Advanced billing pertains to the fixed costs of the bill, e.g. cost of the service 1MR, 1MB, touch tone, directory advertising. With advanced billing you pay the monthly charge for service for the upcoming month "in advance".

Arrears Billing — Customer is billed for service by the telco for the previous month's service. New England Telephone uses Arrears Billing.

BELLCORE — Bell Communications Research. A research company jointly owned by the seven Regional Bell holding companies. BELLCORE sets standards, coordinates USOC implementation and coordinates network services between the Bell telephone companies.

Billing Systems — This is a group of telephone company systems responsible for creating your customer service record (CSR), and rendering your monthly telephone bill.

BOC — Bell Operating Company. One of the twenty two Bell Operating Companies in the United States created by the 1982 Divestiture Agreement.

CABS — Billing system used by the local telephone company to bill the long distance carriers.

Casual Calling — Refers to the ability to use AT&T, MCI or Sprint's long distance network without designating one of them as your primary carrier. For example, to use AT&T you would prefix the number (for interLATA calls) you want to reach with 10288. To use MCI's network, you would prefix the call with 10222, and to use Sprint's network, 10333.

Central Office (CO) — A large telephone switching center which provides dial tone, and routes calls through the telephone company's network. The first three digits of your telephone number indicate the central office that serves your area.

Centrex — Centrex is a business telephone service offered from a local central office. It is basically a single telephone line service to individual phones (the same as you get at your house) with many built in features, including intercom, call forwarding, call transfer, toll restrict, least cost routing and call hold (available on single line phones). Centrex normally comes in two varieties; analog and digital.

Channel Service Unit (CSU) — Channel Service Units are customer provided equipment (CPE) which permits the terminating of digital type service such as T1s.

CO Lines — These are the telephone lines that connect your office or home to your local telephone company's central office. It also connects you to the nationwide telephone switching network.

Collocation — Allows certain telephone company customers to rent space within the telephone company's central offices, and to use this space to set up and maintain their own fiber optic facilities and equipment. This equipment can then be directly connected to the telephone company's network.

Collect Call — A telephone call generally requiring operator-assistance. The called party is verbally asked by the operator if they will pay for the call and if they agree they will be billed.

Competitive Local Carriers (CLCs) — A local carrier which leases capacity on a private fiber network that it has constructed. CLCs provide T1s and other dedicated services in direct competition to the local telephone company. They can also provide limited switched calling services.

Corridor Optional Calling Plan — Offered in certain regions in the United States. In New York, New York Telephone offers a discounted way for subscribers in the 212 and 718 area codes to call the five northern New Jersey counties — Bergen, Essex, Hudson, Passaic and Union for a monthly charge. Using the special access code 10698 (10NYT) calls originating from 212 and 718 and going to these five New Jersey communities can bypass the normal long distance carriers, e.g. AT&T, MCI, etc. The Corridor Optional Calling Plan is also available from the five New Jersey communities to New York City (area codes 212 and 718). New Jersey Bell also bills these calls at a discount. The telephone number dialed would then be preceded by 10658 (10NJB).

CRIS — Customer Record Information System. This is the system that the telephone representative references to retrieve and review the Customer Service Record. CRIS is the system used by the Bell companies to bill residence and business customers.

CSR — Customer Service Record. This is the master record of billing for each customers' account. The CSR lists all the of services and circuits billed on a customers' account. The on line CSR is used by the telco service representative to check your bill whenever an inquiry is made by you.

Demarc Point — The demarc is the point where the telephone company responsibility ends within a customer's premise. This termination point is normally a 42A connecting block or equivalent. The telco maintains the facilities between the demarc and the central office.

Digital Service Unit (DSU) — DSUs are customer provided equipment (CPE) which, along with Channel Service Units, are necessary for connecting digital T1s to the telephone company's network.

Direct Inward Dialing (DID) — You can directly dial a particular extension within a company by using DID trunks, thereby bypassing the PBX attendant. Each extension is assigned a discrete seven digit number. DIDs only allow inward calls. You cannot dial out on a DID trunk.

Divestiture — On January 8, 1982 AT&T signed a consent degree with the Justice Department, stipulating that on midnight December 30, 1983, AT&T would divest itself of its 22 telephone operating companies. According to the terms of the agreement, the 22 telephone operating companies would be formed into seven regional operating companies of roughly equal size. The terms of the agreement placed various restrictions on AT&T and these new RBOCs. Specifically, the RBOCs are not allowed to provide long distance service, cannot manufacture telephone equipment or provide information services. AT&T was not allowed to provide local telephone service in competition with the RBOCs. (The restriction on providing information services has since been lifted). The following details the Bell System — Post divestiture:

AMERITECH — REGIONAL BELL OPERATING COMPANY (RBOC) FOR THE FOLLOWING BELL OPERATING COMPANIES (BOCs):

ILLINOIS BELL
INDIANA BELL
MICHIGAN BELL
OHIO BELL
WISCONSIN BELL

BELL ATLANTIC — REGIONAL BELL OPERATING COMPANY FOR THE FOLLOWING BOCs:

BELL OF PENNSYLVANIA
C & P TELEPHONE COMPANY OF MARYLAND
C & P TELEPHONE COMPANY OF WEST VIRGINIA
DIAMOND STATE TELEPHONE COMPANY
NEW JERSEY BELL

BELLSOUTH — REGIONAL BELL OPERATING COMPANY FOR THE FOLLOWING BOCs:

SOUTHERN BELL
SOUTH CENTRAL BELL

NYNEX — REGIONAL BELL OPERATING COMPANY FOR THE FOLLOWING BOCs:

NEW ENGLAND TELEPHONE COMPANY
NEW YORK TELEPHONE COMPANY

PACIFIC TELESIS — REGIONAL BELL OPERATING

COMPANY FOR THE FOLLOWING BOCs:

PACIFIC BELL
NEVADA BELL

SOUTHWESTERN BELL TELEPHONE — REGIONAL BELL OPERATING COMPANY FOR THE FOLLOWING BOCs:

SOUTHWESTERN BELL

US WEST— REGIONAL BELL OPERATING COMPANY FOR THE FOLLOWING BOCs:

MOUNTAIN TELEPHONE
NORTHWESTERN BELL
PACIFIC NORTHWEST BELL

DMS — Family of central office switches made by Northern Telecom.

DS0, DS1 — Pronounced "D-S Zero" and D-S One". These are units of transmission as defined by bandwidth. DS-1 is T-1, at 1.544Mbps. A single DS0 is 64 Kbps. Twenty-four DS0s (24 x 64Kbps) equals one DS1.

Econopath Calling Plans — New York Telephone's version of an intraLATA economy calling plan available for businesses to make calls within a regional calling area (LATA). For example in New York, Econopath offers Manhattan businesses a discount for calls made to the East Suffolk region.

ESS — A family of central office switches made by AT&T, e.g. ESS 1 and ESS 5.

Exchange Service — Normally associated with a 1MB (Measured Business Line) or 1MR (Measured Residence Line) service which is provided by a telephone circuit (line) terminating at the telephone central office. This line provides dial tone and connects you to the public switched network.

Facilities-Based Carriers — Telephone companies that have their own physical networks, as opposed to companies that resell services provided on other networks.

FACS — Facility Assignment and Control System. This system is used to provision customer line orders.

Federal Communications Commission (FCC) — The FCC regulates all interstate telephone traffic.

Fiber Optics — These are thin strands of glass through which light is transmitted. Fiber Optics offers high bandwidth, low cost, and small space needs.

Field Identifier (FID) — Used by the telephone company to further describe USOCs.

Flexpath — New York Telephone service that provides 1.544 Mbps service between a digital PBX and the serving central office. Flexpath has Direct Inward Calling (DID) and Direct Outward Dialing (DOD).

Gold Number — Also called vanity number. It is a telephone that is easy for your customers to remember, e.g. 767-1111. You pay an additional monthly charge for this number.

Intellipath Digital Centrex Service — New York Telephones' version of digital centrex service. This is Centrex that is provisioned out of a digital CO.

LATA (Local Access and Transport Area) — One of 161 local telephone service areas in the US. As a result of the Bell System Divestiture Agreement, switched calls with both end points within a LATA (also referred as intraLATA calls) are generally the sole responsibility of the local telephone company, while calls leaving the LATA (also referred as interLATA calls) are passed to an interexchange carrier, e.g. AT&T, MCI, Sprint, etc.

Leased Line — A telephone line going from one place to another. You rent a leased line typically by the month and normally do not pay extra for any usage (number of calls carried over the line). It's exclusively yours to use.

LEC — A Local Exchange Carrier which is either a Bell Operating Company (e.g. New York Telephone) or an independent (e.g. GTE), which provides local telephone services. LECs are sometimes referred to as telephone companies or telcos.

LMOS — Loop Maintenance Operation System.

MDF — Main Distribution Frame.

Month-To-Month Billing — The standard way of paying for telephone service. Some telephone company services now come in "rate stability" packages. This means if you commit to the service for three to five years, you pay less each month.

MTS — Message Toll Service. This is regular telephone service, as distinguished from WATS service. It is "full fare" long distance service.

One Way Trunks — These trunks are ordered from the telephone company and can be used to receive incoming calls from the central office or for outgoing service to the central office, but not for two way calling.

POP — Point of Presence. A long distance carriers connecting point with a local telephone company.

POTS — Another name for Plain Old Telephone Service. Just dial tone, no fancy features.

Private Branch Exchange (PBX) — A private (i.e. you, as opposed to the phone company owns it), branch (meaning it is a small phone company central office), exchange (a central office was originally called a public exchange, or simply a exchange). A PBX or smaller version called a key telephone system allows you to connect extensions to the PBX and allows for dialing from extension (station) to extension (station) without the need to involve the telephone company's central office. The PBX provides dial tone, talking battery and ringing generator in the same way the telephone company's central office does for phones directly connected to it. The PBX is in turn connected to the central office by way of central office (CO) trunks (lines) that allow an extension to make and receive calls outside of the PBX.

Private Line (PL) — A direct line or channel between two or more locations that are exclusively available to a particular customer 24 hours a day. This type of line does not have access to the switched network.

Provisioning Systems — This is a group of telephone company systems that are tightly coupled in that they process the service order request that is originated by the Service Order Processor (SOP) system. The provisioning systems assign cable and pair facilities as well as direct central office and field personnel to perform the physical work to satisfy the customers' request as described on the service order. After the work is completed the service order is so noted and the order is then sent to the billing system.

Rate Stability Plan (RSP) — Commit to a term plan for local services for three to five years and you will receive a discount off the regualr rate. The customer is normally insulated from any rate increases occurring during the course plan. Good plan to consider, but beware of penalties if you move during the term period

Remote Call Forwarding (RCF) — Telephone company service that instantly forwards a call to another prearranged number. RCF is a cheap way to establish a "presence" in a distant city.

Service Order — A formal request issued by a telephone company's representative that details the type and cost of telephone service requested by a particular customer. Your service request is translated into USOC format and is the entered into the telephone company's Service Order System (SOP) for provisioning and billing.

Service Order Processor System (SOP) — Each telephone company has an equivalent service order system that is used to enter and process customer requests for new service or any change(s) to existing service. A service order is issued each time you request a change to your service.

SMDR Port — Modern PBXs and some key systems have an Station Message Detail Recording (SMDR) electrical plug, usually a RS-232C, 25 pin connector. The SMDR port is used by call accounting systems and some toll fraud detection units. The telephone system sends information (normally referred to as Call Detail Records) on each call that is made and received. This record normally contains the extension that originated the call, the trunk that handled the call, duration of the call, time and date stamp and pricing information.

Summary Billing — Telephone company service that itemizes the charges for all of your different locations onto one bill.

SUPERPATH — New York Telephone's point to point T1 service.

Switchway — New York Telephone's switched service that lets you dial up and transmit data at 56Kbps. It is end to end digital service.

T-1 — Also spelled T1. It is a digital transmission line with a capacity of 1.544 Mbps (1,544,000 bits per second). A T-1 uses two pairs of twisted wires, the same as you would find in your home. A T-1 normally can handle 24 simultaneous voice conversations with each digitized at 64Kbps. With new voice encoding techniques, it can sometimes handle even more voice conversations. T-1 links can be connected directly to new digital PBXs.

Telephone Exchange — A telephone company switching center for connecting and switching phone lines. A European term for what North Americans call a central office.

Third Party Call — Any call charged to a number other than that of the origination or destination party.

Tie Line — A dedicated circuit that connects two PBXs.

Tip & Ring — An old fashioned way to refer to twisted wire, but the term is still used throughout the telephone industry. Ring is normally the red wire and the tip is green wire with ground being yellow wire. Tip also refers to the transmit side and Ring to the receive side of a circuit.

TIRKS — Trunk Integrated Record keeping System.

Tone Dialing — A push button phone that emits a different sound (frequency) for each digit or special function appearing on the touch tone key pad.

Touchtone — A trademark owned by AT&T for tone dialing.

Trunk — A communications line between two switching systems. A central office (CO) trunk connects the customers PBX switch to the telephone company local serving central office. A tie trunk (or line) connects two or more PBXs together for passing voice and data between them.

Twisted Pair — Two insulated cooper wires twisted around each other over the length of the cable to reduce induction noise. Twisted pair is also synonymous with local loop or subscriber loop used to connect the local subscribers equipment (e.g. PBX, telephone set, etc.) to the local serving central office.

Two Way Trunk — A trunk that is used for two way conversations into or out of a telephone system, i.e. most trunks. Some trunks are used as one way for either receiving calls or making calls but not both.

Usage Based — Refers to a rate or price for telephone service based on usage rather than a fixed monthly rate. An exchange line is charged a usage rate based on number of call made and also for the duration of each call.

User Loop — A 2 or 4 wire circuit that connects a user to the central office.

USOC — Universal Service Order Code. (Pronounced "U-Sock"). A Bell System term identifying a particular service or equipment offered under a tariff. The USOC code is a by product of the old Bell System where AT&T set standards for all of the 22 Bell Operating Companies. Since divestiture BELLCORE is responsible for setting standards for the BOCs. The idea behind using USOCs is to identify every service or product offered by a telephone company with a discrete code that can be read and interpreted by sophisticated service order and billing systems. There are over 30,000 USOCs in use by the Bell Operating Companies.

V & H — Vertical and Horizontal grid coordinates. The V&H numbers are used to determine "airline distance" between rate centers (central offices). Each central office is identified by V&H coordinates which are used to calculated mileage billing on leased lines and other mileage sensitive services.

Voice Circuit — A typical analog telephone channel coming into your house or office. It has a bandwidth between 300Hz and 3000 Hertz which is sufficient to recognize and understand the person on the other end of the circuit.

VPN — Virtual Private Network. A service offered by the long distance carriers to large companies that wish to link multiple locations without having dedicated physical lines between those locations.

WATS — Wide Area Telecommunications Service. Basically discounted toll service provided by all long distance and local telephone companies. AT&T started WATS but forgot to trademark the name, so now every supplier uses it as a generic name. There are two types of WATS services; in and out WATS, i.e. those WATS lines that allow you to dial out and those on which you receive incoming calls.

Yellow Pages — A directory of telephone numbers classified by type of business. It was printed on yellow paper throughout the 20th century.

APPENDIX A

TOLL FRAUD/TOLL ABUSE LOSS PREVENTION: AUDITING YOUR VOICE NETWORK SECURIITY MEASURES

PBX TOLL FRAUD WHAT IS IT?

Remote PBX toll fraud is the unauthorized use of your voice network by individuals located geographically outside your premises. These unauthorized users call themselves hackers. It is estimated that businesses lose over 3 billion dollars a year to PBX toll fraud.

Hackers can break in to your PBX via DISA (Direct Inward System Access), remote maintenance ports and through your voice mail system. The FCC has ruled that the customer, not the long distance carrier is responsible for payment of the fraudulent calls (Chartways vs. AT&T). With fraudulent calling card calls you are only responsible for the first $50 of usage.

Toll abuse refers to unauthorized calls made by employees or by visitors to your firm. It includes calls to 900, 976 and 700 series numbers, and excessive personal calls. An employee making frequent personnel calls is not doing the work that he or she is being paid to do.

WAYS TO REDUCE YOUR EXPOSURE TO TOLL FRAUD

The responsible telecom manager can take steps to reduce all aspects of toll fraud and toll abuse. You should review and document all security measures taken by your company. Most importantly, they should be monitored for effectiveness. Your audit should ensure the following steps have been taken to protect your company's voice network.

Step 1 - Perform an audit of all your telecommunications facilities. Remember, you can not protect telephone lines that are not documented. Disconnect all unused POTS and special service lines. Make sure that all existing telephone facilities are assigned to a particular person or department.

Step 2 - Prepare a network diagram of all your facilities. Your network security is only as good as the weakest node connected to it. If, for example, you have TIE lines connecting PBXs at different locations, you must secure all locations. If one PBX is compromised then all your locations are at risk.

Step 3 - Assign a responsible department to all telephone facilities. Each department should be given a list of facilities that it is responsible for. Each department should also be required to review its telephone facility list to ensure that all facilities are required. Each department then becomes responsible to notify the telecommunications department when they down size and no longer need certain facilities.

Step 4 - Monitoring. Your basic defense against toll fraud/toll abuse is the use of a call accounting system. (If you do not have a call accounting system, install one ASAP). Call accounting reports should be checked daily for unusual calling activities. State of the art call accounting systems come equipped with alarm generators. (Toll fraud monitoring devices can also be purchased and installed independently of your call accounting system).

You can set alarms to detect unusual calling patterns. This is done by setting parameters that define acceptable calling patterns. You can set a parameter, for example, that tells your call accounting system to set off an alarm when an over seas call from a particular extension exceeds 20 minutes. Alarms can be sent to PCs or pagers to alert you to abuse. Alarms can also be sent to central monitoring stations that act like burglar companies.

Step 5 - Tracking. You must require all departments to budget and estimate monthly telephone costs. Actual totals should be compared to estimates. Reports should be generated that track trends by department. Telecommunications is an expense that should be borne by the user of the services.

Step 6 - Blocking by the local telephone company.
A. You can block 900, 700 and 500 series calls through your local telephone company. They will charge you a one time fee to block these calls (usually about $25 to $30). Lines that have blocking can be identified by checking your CSR. A New York Telephone CSR will identify blocking as follows:

QUANT	ITEM	DESCRIPTION
1	RTVXC	/LCC ABB (BLOCKING SERVICE CHARGE)

B. You can also block third number calls and collect calls on each individual telephone line number. On a New York Telephone CSR TBE A indicates that both collect and third number calls are blocked on a particular line. TBE B indicates that third number calls are blocked on a particular line while TBE C indicates that collect calls are blocked on a particular line. An example of how this will appear on a New York Telephone CSR is as follows:

QUANT	ITEM	DESCRIPTION
1	TXG	74,TKNA,718,999,1234 /TBE A/PIC AT&T

C. A line can have call screening placed on it. Call screening is typically utilized by hospitals and hotels. It prevents you from making a directly dialed call from a particular line. An operator will intercept all calls and force you to place the call collect or bill it to a credit card.

Step 7 - Blocking by the long distance telephone company. You can have your long distance carrier assign access codes to particular line numbers. They can also be assigned by department or to each employee. Certain access codes can be restricted from dialing certain area codes. This type of blocking only works for calls placed over your primary carrier's network. If employees or visitors dial the access code for other than your primary carrier, they will circumvent this blocking.

Step 8 - Blocking at the PBX. Your PBX can block access to certain area codes by telephone line or extension by what is called class of service assignments. Certain lines can be restricted while others are left unrestricted. You can also restrict by time of day. For example you can block all calls made from a telephone line after 6 P.M. Calls to high risk area codes (i.e. 809) should be blocked where possible.

Step 9 - Institute Voicemail Security Measures.
A. Block all forward to the public network features on your voicemail.
B. Change Voicemail passwords quarterly.
C. Delete unassigned voicemail boxes.
D. Make sure passwords exceed 4 characters and are alphanumeric.

Step 10 - All remote maintenance ports and inbound modem lines should be disconnected when not in use or they should be equipped with call back devices.

Step 11 - Disconnect or restrict DISA access. The most common way hackers compromise your voice network is breaking your DISA passwords.

Step 12 - Institute system access password security measures.
A. Make sure your initial PBX default password has been changed.
B. Change PBX access passwords quarterly.
C. Delete access codes for employees that leave the company ASAP.

Step 13 - Prepare a Toll Fraud Prevention Plan. Document and review all prevention measures taken to protect your voice network. Update this plan as the business changes.

Step 14 -Prepare a Toll Fraud Contingency Plan. This plan documents the steps you will take if PBX toll fraud is detected. It will detail the steps you will take if a hacker has successfully broken into your voice network. This plan should include the 24 hour numbers of your long distance carrier, maintenance vendor and the home number of three responsible employees. You should also obtain the 24 hour network security number for AT&T, MCI and SPRINT. This companies allow casual calling, and hackers typically place fraudulent calls over a variety of long distance carriers.

TOLL FRAUD INSURANCE

Your ultimate protection from toll fraud is an insurance policy. This industry is still in its infancy and policies are limited and expensive. Until a history of claims is established, insurance carriers will be reluctant to offer toll fraud insurance. Your choices are currently limited to either "insurance like" offerings from SPRINT and AT&T or the Travelers' "Remote Access Telephone Fraud Policy". Aetna has a toll fraud policy, but does not market the policy to the general public.

SPRINT calls its toll fraud insurance service offering SPRINTGUARD, while AT&T calls its offering NETPROTECT. Both programs have inherent conflict of interest problems. If a claim of toll fraud is made they become both the company that bills you for fraudulent calls (and expects payment) and the company that must decide if you are covered under their insurance plan.

SPRINTGUARD

"THE PRO's"

Cost. SPRINTGUARD's monthly fee is relatively low cost. Their fee structure is set as follows:
One time activation fee:
$100 per PBX
50 or more PBXs - $5,000
Monthly maintenance:
$100 per PBX
50 or more PBXs - $5,000

THE CON's"

To qualify for SPRINTGUARD PLUS you have to sign a 2 year contract with Sprint and commit a minimum of $30,000 in voice services per month to Sprint.

Limitations:

A. Does not cover toll fraud that originates from your PBX but occurs over another long distance carrier's network. Most hackers today, once in, will switch carriers to avoid detection. They simply prefix the fraudulent calls with the access codes for AT&T (10288) or MCI (10222).

B. Only covers fraud calls on Sprint's dedicated service offerings (Ultra WATS, VPN, and dedicated Clarity). Switched calls, routed by the local telephone company, are not covered.

C. Does not cover fraud that terminates within the United States, Puerto Rico, or the Virgin Islands.

SPRINTGUARD PLUS comes with a deductible of $25,000 and a liability limit of 1 million dollars.

Conclusion - Sprint is using this offering to appear proactive on the issue of toll fraud. In fact, the offering is designed to get customers to commit to SPRINT's dedicated long distance service for extended periods.

AT&T's NETPROTECT Advanced service is very similar to SPRINT's SPRINTGUARD service. To qualify for NETPROTECT Advanced a customer must meet the following conditions:
Customer must have both AT&T 800 service and AT&T outbound long distance traffic at each covered site.
Customer must commit to service for a twelve month period.
Limitations:
Only covers fraud on AT&T's network.

Once fraud is detected by AT&T the client is notified and is given two hours to correct the situation. If the fraud continues after the two hour warning then the subscriber becomes fully liable for all subsequent fraud.

Policy has a $25,000 deductible and a cap of one million dollars. The monthly and establishment charges for NETPROTECT are as follows:

Service establishment charge per PBX- $120
Monthly charge per PBX:
1-50..$120
51-100..$100
101-150..$80
151-200..$50
over 201..$25

AT&T also offers a deluxe version of NETPROTECT Advanced called NETPROTECT Premium. NETPROTECT Premium eliminates the deductible but raises the monthly and one time establishment charge. NETPROTECT Premium charges the following:

Service establishment charge per PBX-$300
Monthly charge per PBX:
1-50..$225
51-100..$200
101-150..$175
151-200..$150
over 200..$100

Travelers

Travelers announced toll fraud insurance on August 21,1992. It covers fraud that occurs on any long distance carrier network. It is, however, very expensive. Under the Travelers' plan, annual liability limits range from $50,000 to $1 million with deductibles set at 10% of the liability limit. Premiums range from $2500 to $49,000. For example, a $50,000 coverage limit with a deductible of $5000 will cost approximately $2500 and a $1 million coverage limit with a deductible of $100,000 will cost you approximately $50,000. Discounts are available if you install a toll fraud monitoring device or other security devices.

PRO's
Covers fraud that occurs on any carrier's network

CON's
The premiums are high for the deductibles and limits offered.

AETNA

Aetna Casualty and Surety Company has been offering a form of toll fraud insurance since March 1992 to banks and S&Ls. To obtain the coverage a company must purchase a bond that covers computer theft and internal fraud. Aetna will cover up to 5 million dollars per incident, but does not release specific prices as all policies are customized.

Other insurance companies are working on toll fraud policies. Premiums should decrease as the competition increases. The industry will be better able to price toll fraud insurance as a database is developed that tracks toll fraud claims versus revenues.
Toll fraud activity will continue to increase. No program or service can offer you 100% protection. The smart manager will stay vigilant and constantly review and update his Toll Fraud Prevention and Detection Contingency Plan.

APPENDIX B

10 STEPS THAT THE BELL OPERATING COMPANIES CAN TAKE TO IMPROVE CUSTOMER SERVICE AND BILLING ACCURACY

1. All Customer Service Records (CSRs) should be printed on clean, columned paper. They should not be local prints of computer terminal screens. (Southern Bell and the New York Telephone Company receive high marks, while New Jersey Bell rates low in this category).

2. Customers should be educated and encouraged to review and obtain their CSR. Currently, most of the BOCs provide a general breakdown of the monthly billing charges once a year. This breakdown does not provide enough information to the customer, and in fact gives the customer a false sense of security. Many customers do not realize the amount of information that is only available to them on the CSR.

3. CSRs should have an audit trail of activity for each USOC billed on a customer CSR. Currently, each time a rate change occurs, the activity date is overlaid with the rate change date. This confuses both customers and many telephone company representatives, as they erroneously believe activity (meaning physical activity) has occurred on a particular line or circuit on a particular date. Charges for lines or circuits are often not challenged as it seems physical activity is taking place on these services.

3. USOC information should be disseminated to large customers and consultants on request. (Pacific Bell and New Jersey Bell receive high marks, while New York Telephone receives low marks in this catergory).

4. Special service charges should be itemized on a separate and distinct bill. Most Bell Companies intermix charges for special service circuits with charges for POTS lines. When charges are intermixed, bills are hard to verify. Many companies have seperate voice and data departments. Seperate billing by the telephone company will help these companies assign verification and payment responsibilities to the proper department. (Pacific Bell receives high marks in this category while New Jersey Bell and New York Telephone receive low marks).

5. Provide customers with the option of receiving bills via EDI (Electronic Data Interchange) free of charge. Customers should also be able to pay bills via EDI. Interim methods like New York Telephone's Summary Billing, which is provided by magnetic tape, is actually quite costly to the customer.

6. Rate quotes should be available to customers via a computerized database. Customers should have on-line access to telephone company tariffs. AT&T is particularly proactive in providing inter-LATA rates for its products and services. For example, AT&T provides an 800 number for toll free access to a electronic library that details product and rate information.

7. The local telephone companies should be required to have the customer to sign an authorization form before they bill wire maintenance charges. These charges mysteriously appear on many bills, and often cause confusion and resentment when customers discover them. Most telephone companies can not tell you how some of these charges got onto their bills.

8. Increase the use of independent sales agents, but exercise increased surveillance of their methods and procedures.

9. Instruct Customer Service Representatives not to use early retirement and/or layoffs as excuses for delays. With most of the local telephone companies reducing their work force, customers are often told by representatives that they are shorthanded. A customer with a problem does not want to hear that the phone company is shorthanded.

10. Telephone companies should start internal auditing units that randomly verifies billing (free of charge to the customer). They should be as knowledgeable and motivated as independent auditing firms. These units should also market themselves to customers and offer to provide applications engineering services to interested firms. The good will generated and the feedback obtained from these audits will prove invaluable to increasing customer loyalty and billing accuracy.